"... I have always loved it — the good smell of the pines, the sound of the axes and the crash of falling trees, the creak of the big teams swinging down the mountainside, and through it all an air like wine and a sparkling sunlight."

*Ruth Park, 1916*

Other Books by
## Moose Country Press

### Jack Noon
The Big Fish of Barston Falls (1995)
Old Sam's Thunder (1998)
The Bassing of New Hampshire (1999)
Up Moosilauke (2000)

(Writing as M.J. Beagle)
Sit Free or Die (1994)
The New Hampshire Primer (1995)

### William S. Morse
A Country Life (1995)
A Mix of Years (1998)

### Robert W. Averill
The Moosilaukee Reader (1999), Volumes I, II

❄ ❄ ❄

# LUMBER QUEEN

The Life of Woodswoman
Ruth Ayer Park

by Ellen C. Anderson

Moose Country Press
2001

Moose Country Press
Warner, N.H.

© 2001 Ellen C. Anderson

All Rights Reserved

*No part of this book may be reproduced in any form or by any electronic or mechanical means, including information storage and retrieval systems, without permission in writing from the author, except by a reviewer who may use brief parts in a review.*

ISBN 1-893863-01-8

**Library of Congress Cataloging-in-Publication Data**

Anderson, Ellen C. (Ellen Clark), 1946–
    Lumber queen: the life of woodswoman, Ruth Ayer Park / by Ellen C. Anderson.
    p. cm.

    ISBN 1-893863-01-8 (pbk. : alk. paper).
    1. Park, Ruth Ayer     2. Lumber trade -- New Hampshire
    3. Logging -- New Hampshire -- History.
    4. Loggers -- United States -- Biography. I. Title.
HD9759.P37 A53    2001
338.7'674'092 – dc21
[B]                                                             2001045028

10 9 8 7 6 5 4 3 2 1
Printed in the United States of America

This book is dedicated to

Rhoda Shaw Clark
Elizabeth Yates McGreal
J. Willcox Brown
Lyle Moody
Joseph D. Park

\* \* \*

Ruth Ayer Park
*circa 1900*

# FOREWORD
## J. Willcox Brown

It is a rare occasion when a personality recognized as unique in her time is brought vividly to life by a writer in a subsequent period. Ellen Anderson has given us a warm and rounded view of Ruth Ayer Park not only in her distinctive role as a lady logger but also in her quiet appreciation of nature and the arts.

Different trajectories on Mt. Moosilauke eventually brought me within Ruth Park's orbit. At first it was an unintended approach when I led the trail crew from the mountain to Old Home Day in Warren in 1936. Fred Gleason had on display his remarkable photos, including logging on Moosilauke at the turn of the century. When I met with Fred later in 1939, I learned his mountain scenes were of the W.R. Park logging operation.

My ties with the town of Warren had been strengthened during construction of the new Ravine Lodge, under the masterful leadership of Ross McKenney and a skilled local crew. After its completion in 1939, I left Dartmouth as manager of the Outing Club and headed for Yale Forestry School, still unaware of the redoubtable Miss Park.

In New Haven I requested as a thesis topic the forest history of Mt. Moosilauke. Being skeptical of time and distance, my tough old professor said: "You can try." Thus I was committed to frequent nostalgic visits to "Dartmouth's Mountain" with less than two years till graduation. Staying awake in Monday morning classes was the punctuation of the challenge.

Pursuing the lead of the Gleason photos, I was soon pointed in the direction of the survivor of the Park operation, Ruth Ayer Park.

Blessed with a precise memory and a penetrating perception of the human scene, she was my chief source of the history of W.R. Park's Gravity Railroad. It had traversed the rugged terrain from Camp #2 (later the original Ravine Camp) to the impassable gorge at the iron bridge site a century before this millennium. Refraining from personal inquiries, I did not know then that Ruth would have been in her mid-teens at the time and obviously a keen observer of her father's engineering exploit.

Later she personally performed a direct role with the Park land on Moosilauke, which lay on the east slope and extended high on the main mass of the mountain. Much of the tract had been logged substantially. After years of trying, Ruth found the right buyer in the Parker-Young Company after they acquired the other half of Jobildunc Ravine. Ruth's triumph of salesmanship involved skill on snowshoes and amazing familiarity with the rugged headwaters of the Baker River (which many of us prefer to term the Asquamchumauke). The price was right, and, to Ruth's delight, made possible the settlement of her father's estate.

Since my first exposure to the White Mountains was not until 1927 as a twelve year old summer camper from Delaware, Ruth Park's direct involvement with Moosilauke was concluded before my initial ascent.

Thus I did not meet Ruth Park at Moosilauke or in Warren, but on the Park family property in isolated Dorchester. There the Parks still owned a substantial forest and the Maskoma Lodge summer camp on Cummins Pond. The camp was run by Ruth's youngest sister, Katherine. It must have been as hike leader and teller of tales that Ruth became "Auntie." In any event this diminutive dynamo was generally known as "Auntie Ruth" by the time I met her. That she made a profound impression on the summer campers, I learned from Delaware friends who attended the family camp.

It was in her off-season role as logging boss that this tiny lady, thirty years my senior left the keenest recollections with me. A remote logging camp, often snowbound, was scarcely the place you'd expect to find in the charge of a Vassar graduate with a taste for fine literature and classical music. Though her quarters in the Lodge were separate from the logging camp, it would have been far too precarious a situation for any woman less determined than "Auntie Ruth."

In 1941, I invited Natale Linton, my future wife, to be my Carnival guest. While in Hanover, I suggested that we visit the logging boss at Cummins Pond. The reception there was kindly but the access almost disastrous. Beyond Lyme Center the dirt road's winter maintenance was chiefly by log trucks. Driving a well-amortized sedan, I struggled unsuccessfully to stay out of the deep ruts and became high-centered on the icy ridge. With a double-bitted axe in the trunk, I managed to chop us free.

So "Auntie Ruth" bestowed her own benediction of treating us like fellow survivors!

**Will Brown**

# Contents

*Foreword — J. Willcox Brown*
*Author's Preface*
*Acknowledgements*

| | |
|---|---:|
| I. Prelude | 1 |
| II. An Historical Fiction of the Childhood and Education of Ruth Ayer Park | 19 |
| Chapter 1: Childhood Affections, Plymouth, 1890 | 21 |
| Chapter 2: Early Lessons | 27 |
| Chapter 3: Back to Warren | 33 |
| Chapter 4: The Family Gathers | 39 |
| Chapter 5: Timber Pirates | 45 |
| Chapter 6: Safe at Home | 51 |
| Chapter 7: Logging Days | 55 |
| Chapter 8: New Beginnings | 59 |
| Chapter 9: Modern Methods on the Farm | 65 |
| Chapter 10: Pacific Vistas | 71 |
| Chapter 11: Now or Never | 75 |
| Chapter 12: High Gear | 83 |
| Chapter 13: Retreat | 87 |
| Chapter 14: Home Free | 93 |
| III. Lumberjack Tales | 101 |
| IV. Letters and Other Communications | 117 |
| V. Postlude: Factual Account and Commentary | 133 |

# Author's Preface

*"She lives alone in the woods in a cabin without electricity."*

I recall as a child my mother telling me about Ruth Park, who was a friend of my grandparents, had graduated from Vassar, and ran her father's timber business. Family stories persisted of Ruth's colorful life. I thought, "How strange. Why would anyone want to live that way?"

In November, 1992, my husband and I moved to the Baker River Valley in New Hampshire, the area where Ruth Park had lived from 1885 to 1980. Mother was visiting and suggested, "Let's go find where Ruth Park lived!" We spent an afternoon exploring Warren, but the exact spot where the Park farm had stood eluded us, and Mother went home, still curious. I resolved to research "just a few facts" about Ruth's life to give Mom as a Christmas present. Local residents were forthcoming with details once they learned Ruth had been a friend of the family and the account began to intrigue me.

I HAD TO KNOW more about Ruth Park! I enlisted Mother's help and found I couldn't stop until I had the whole story. Then I began to write.

I've organized the book into five parts:

**Part One,** *Prelude,* is about Ruth Park's family, especially about her father, W.R. Park, and the activities of the Park Lumber Company in northern New Hampshire.

**Part Two,** *The Childhood and Education of Ruth Ayer Park,* takes all the facts and fills in where there is no recorded information, to create an "historical fiction" of the Lumber Queen's life. This section tells the story of how it might have been, growing up as the daughter of a "timber baron." It is suitable for reading aloud to children, or for younger readers, but certainly it is enjoyable material for all.

**Part Three,** *Lumberjack Tales,* was written by the Lumber Queen herself, providing a never-before published collection of narratives as told by two of the woodsmen who worked in Ruth's lumber camps.

**Part Four,** *Letters and Communications,* gives us a glimpse of Ruth's perspectives on life as we read her letters to friends and family.

**Part Five,** *Postlude,* returns to the facts and finishes the account, including recollections of those who were acquainted with the Lumber Queen.

Ellen C. Anderson
Rumney, New Hampshire
April, 2001

# Acknowledgements

**Special thanks** are due to the following individuals and organizations.

Karen A. Anderson
Paul T. Anderson
Cynthia Aguilar
Nella Ashley
Janet Baker
Alfred Balch
Clifford Ball
Leslie Ball
Robert Bancroft
Roland Bixby
Edwin and Katharine Blaisdell
Ruth Goodhue Boyer
Lilla McLane Bradley
Lloyd Cate
Richard Chisolm
Alexander S. Clark
Winfield S. Clark
Elizabeth Daniels
Roger Daniels
Louis Delsart
Libby Thayer Drowne
Cora Emlen
Claude Foote
Wistar Goodhue
Edith Gray
Bob Green
Ken Haedrich
Elmer Heath
Eudora Hibbard
Harold Hildreth
Raymond Keniston
William Laffer
Elsie Hill – Levitt Latham
Richard Learned
Harold and Thelma MacDonald
Nancy McKechnie

*continued . . . .*

# Acknowledgements
*continued*

Helen and Jim Nichols
David C. Nutt
Jeff O'Connor
Zola Ostrander
Ginny Park
Maria Park
Dana Philbrook
Isabel Pushee
Malcolm Ray
Everett Rich
Ruth and Patsy Sargent
Leon Sharp
Aaron Shortt
Mrs. Norman Smith
Jay Sobetzer
Virginia Harriman Spead
Wilfred Tatham
Robert Thayer
Elsa Turmelle
A.M. Veazey
John Veazey
Henry Waldo
Esther Whitcher
Eugene Whitcher

Burt Russell Litho – Gary Russell, Brian Niedbala
Vassar College Special Collections

Dorchester Historical Society
Plymouth Historical Society
Rumney Historical Society
Warren Historical Society
Wentworth Historical Society

*This book was prepared for publication by
Robert W. Averill and Jack Noon of Moose Country Press.*

# PART I

## Prelude

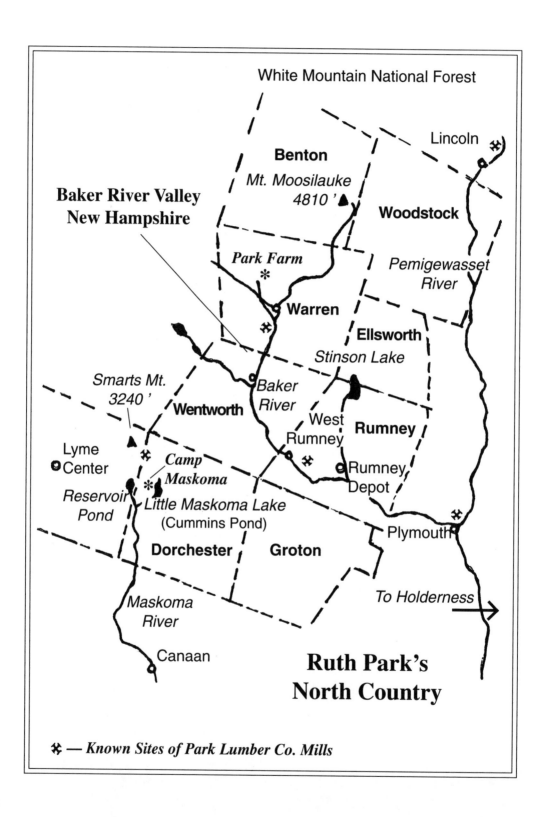

# PRELUDE
## Ruth Park's Family History

The pleasant vistas afforded by the convergence of two river valleys make the region around Plymouth, New Hampshire, a welcome sight to travelers from any direction. Out of the north, from the White Mountains, comes the Pemigewasset River. From the west flows its smaller cousin, the Baker. One is struck by the beauty of the intervals, gentle plateaus of river-bottom land, formed by the changing course of water and ice over time. Abundant forests blanket the mountainsides, which slope abruptly to the valley floors. Anciently, Native American tribes traveled the region on their seasonal round-trips from Canada to the Atlantic Ocean. The rivers they camped along teemed with trout and salmon. Large clay deposits in the riverbanks inspired extensive pottery making.

In the late 1700s, settlers of European backgrounds arrived: first, explorers and trappers followed soon by farming families. The intervals made excellent cropland, and less than a century later, 150,000 acres had been cleared in the Baker Valley alone.

The farmers in the towns of Rumney, Groton, Dorchester, Wentworth, and Warren, as well as Plymouth, were developing goods for market — wool, dairy products, cider, maple sugar, and potatoes. Turnpikes thundered with the hooves of horses drawing buckboards, freight wagons, pungs, and coaches, loaded with products, supplies, mail, and people. The colorful drivers, known as Knights of the Whip, brought stories from afar, and they told of another thundering horse, the Iron Horse, the magnificent railroad engines steaming throughout America. Baker Valley merchants urged the building of a railroad line that would connect Montreal to Boston, via Plymouth.

"The first train steamed into Plymouth January 18, 1850, with over one hundred tons of merchandise and as many passengers," writes Joseph Bixby Hoyt in *Baker River Towns*. "The impact of this new transportation system on the valley economy was great . . . . [F]ield crops doubled in value . . . . [S]pruce and hemlock boards that in 1820 sold for three to four dollars per thousand board feet were now worth three times as much; clear pine formerly worth five dollars was now worth twenty-five dollars."

This vibrant economy attracted many newcomers. Two families interest us: the Dodges, who arrived because of the railroad, and the Parks, because of the timber. Their histories, like the two rivers, flow together in Plymouth.

**Joseph A. Dodge**
*(1818-1883)*
Ruth's maternal grandfather as a young man.

For generations, the Dodges farmed in New Boston, a town eighty miles south of Plymouth. Joseph A. Dodge, however, tried school teaching until he became totally intrigued with the railroad industry. He hired on at the Boston and Lowell Railroad as a clerk, and married his hometown sweetheart, Mary Tewksbury, in 1843. Joseph was promoted to be the station agent in Tilton, then Meredith, and finally in Plymouth. *The Dodge Genealogy of Sussex, Massachusetts* reports: "Here was an opportunity for the growth of a man along with the growth of a railroad, and Mr. Dodge held various positions until he became successively superintendent, director, and finally, general director. He was particularly interested in the development of the White Mountain Region, and he early saw the possibilities of the Weirs (on Lake Winnipesaukee) as a popular summer resort."

When Joseph became the general manager of the system in 1852, he was demonstrating his fraternity in a new class of managers, the middle-class professional. Thomas J. Schlereth writes about the railroad industry in his book, *Victorian America — Transformations in Everyday Life*. "[A]n economy of scale developed, since a large railroad operation demanded the daily coordination of hundreds of workers — across thousands of miles of track . . . [comparable] more to maintaining a large army than to running a family business." Schlereth points out that several railroads had former Civil War generals as presidents. Richard Learned of Rumney, a second generation trainman, recalls his father's remarks that Joseph Dodge commanded respect, and was admired for his ability to remember names of employees he'd worked with years before down-line.

Joseph and Mary Dodge had built for themselves a four-chimney brick and clapboard manse on South Main Street. Here they raised a son, John, and a daughter, Elizabeth Andrews Dodge. Elizabeth matured to mirror the confidence of her father as well as the community spirit of her mother. Sixteen female students, inspired by a Plymouth teacher, banded together to form the Young Ladies Library Association, and Elizabeth was one of these students. When Wellesley College opened its doors in 1875, Elizabeth numbered among the three hundred women enrolled. For two years, she participated in the rigorous schedule of academics, and established a lifelong proficiency at singing and drawing.

Records do not yet reveal why she didn't complete the four years at Wellesley. In 1877, Elizabeth's brother John died. Her father's health was declining. However, facts also point to a

Home of Ruth's grandparents, Mary T. and Joseph A. Dodge, on South Main Street, Plymouth, N.H.

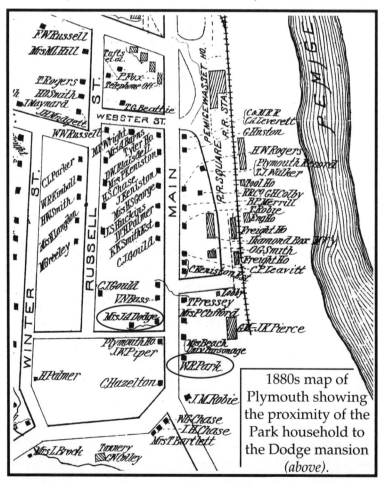

1880s map of Plymouth showing the proximity of the Park household to the Dodge mansion (above).

courtship in Plymouth by a handsome young timber manufacturer, William Richardson Park, Jr.

The Parks came from the Ashby-Townsend area of Massachusetts, just south of the New Hampshire border. The immigrating ancestor was William Park of Glasgow, Scotland, who arrived in America prior to 1799. He married Betsey Wyman, of Ashby, and the couple apparently moved north to Claremont, New Hampshire, where Park acquired a license as an innkeeper. They had two sons, Abel and William. Then, the Scotsman Park disappears from the historic records. (The only trace this author located was in a letter of Ruth Park's, the great-great granddaughter. She indicated that the man had run away in a peddler's cart with a gypsy woman.) Betsey returned to Ashby and eventually remarried a well-to-do landowner named Richard Richardson. The new stepfather must have left a positive impression on the Park stepsons, for the second son, William, upon maturity, married and named his son, William Richardson Park.

When W.R. Park was of age, like his Scotch ancestor, he ventured north to New Hampshire. He settled briefly in Haverhill, along the Connecticut River, where he began a charcoal-burning business. (Charcoal was used by blacksmiths. Enormous quantities were also needed to fire the blast furnace at Franconia Ironworks.)

He married the daughter of Walter Ayer of Haverhill, Lucy Malvina. They began a family, producing two daughters and a son, and moved to the Plymouth area, where he engaged in timber manufacturing. He and Lucy settled the family in the nearby village of Rumney.

All three of the children matured and married prominent citizens. One daughter married a doctor; the other, a business man. The son, William Richardson Park, Jr., married Elizabeth Andrews Dodge on June 20, 1879.

Later that summer, William and Elizabeth moved to Providence, Rhode Island, so that William, or Will, now age 22, could attend Brown University. His course of study included Latin, Greek, French, Trigonometry, Geometry, Algebra, and he graduated *cum laude* in 1882. During these years, two children were born — Mary Elizabeth and Joseph Dodge Park.

Back in Plymouth, Elizabeth's father had been forced to resign as general manager of the B. C. & M. line due to poor health. His physician recommended a retreat to a warmer climate, and in the winter of 1882, Joseph Dodge traveled to Los Angeles for the winter, seeking the cure for his ailments. The real estate of southern California interested him, and he purchased land on 7th Street, and in Pasadena as well. He returned to Plymouth, and passed away on August 10, 1883.

Elizabeth became the heir of his estate. (Probate records mention an adopted daughter named Abbie, as well.) Pregnant, and with two young children, she returned to Plymouth with her husband. They moved into a house on South Main Street, across from Elizabeth's mother. Will rejoined his father in the operation of Park Lumber Company, and reflected a new confidence. He had proven his ability to maintain the rigorous disciplines of Tilton Academy and Brown University. Now, he would apply the same to the timber manufacturing business.

**Joseph A. Dodge**
An important railroad figure in
Plymouth, N.H., from 1850-1883.

The Pemigewasset House on the west bank of the Pemigewasset River in Plymouth was a key Boston, Concord, & Montreal stopover. The train station was attached to the hotel's lobby. Nathaniel Hawthorne, traveling with Franklin Pierce, died at the Pemigewasset House in 1864.

Ruth's father, W.R. Park, Jr., built this steam sawmill in Warren, N.H., at the beginning of the 1900s.
*Photo courtesy of Jeff Belyea*

### History of Park Lumber Company

Park Lumber Company rode the rising tide of the spruce harvest, which began as white pine lumbering reached its peak in 1861. Henry C. Waldo of Lincoln writes, in a paper entitled *Cutting Practices Up to The 1940's*, "The Spruce Operators were forced to look to the high country of the Adirondacks, the Green and White Mountains of Vermont and New Hampshire, and the great Spruce Fir forests in northern Maine." As American townships and cities expanded, the demand for framing timber grew as well, and lumbermen soon "overcame their distaste for the inferior Spruce, which gained wide spread reputation and nothing was allowed to stand in the way of the Spruce harvest."

W.R. and Will Park began operations in Plymouth, north of town, where the Baker converges with the Pemigewasset on a hill later known as Foster's Peg Mill Hill. They established mills in Dorchester, Rumney and West Rumney, and also in Lincoln. Speaking of Lincoln's logging history, Waldo writes, "Lumber manufacture began in 1890 by Mr. W.R. Park, later to become a famed lumberman of Warren. This sawmill, on the East Branch of the Pemi, sold in 1894, just before J.E. Henry and Sons came to town."

Warren, N.H. – Park Lumber Company pond and mill with over a million board feet of spruce from a winter's work.
*Fred Gleason – Esther Whitcher Collection*

Posing (above) with the crew at Park's Warren sawmill is W.D. Veazey, the attorney from Laconia who was appointed by the court during the assigneeship. Below is the Mead & Mason Peg Mill, c. 1894, purchased by W.R. Park for his steam lumber mill, Warren, N.H.
*Esther Whitcher Collection*

W.R. Park Stable, close to the tracks of the Boston, Concord & Montreal Railroad near the Warren town common.
Below: Knights of Pythias Lodge — thought to be the Park Lumber Company store with lodging upstairs for teamsters.

A logging operation, c. 1900, in Warren, N.H.
*Esther Whitcher Collection*

Long spruce logs on Main St. Warren, N.H.
*Esther Whitcher Collection*

Employees of Park Lumber Co. at the Warren sawmill (1904).
*Esther Whitcher Collection*

Concurrent with the building expansion, paper mills were producing more and better types of paper for a growing industrial world. The digesters that produced the paper pulp constantly needed supplies of wood, as much of the paper was made from wood fibers. But the small-scale operations of the past were not able to satisfy the demand. New Hampshire forester and historian J. Willcox Brown, in his popular work, *Forest History of Mount Moosilauke*, explains, "The unequal opportunity of the small operator to compete was based chiefly on transportation costs and lack of capital for large-scale facilities."

Stronger, more professional companies emerged, and Park Lumber was one of those ready to make the adjustment to a larger-scale operation.

Experience, desire, and capital were on their side. Twenty years of managing crews and sawmills had provided practical training for W.R. and Will. Young Will wanted to provide a lifestyle that Elizabeth and the children could be proud of, and at the same time, preserve the respect that he and his father enjoyed as entrepreneurs. The strength of this desire captivated Elizabeth Dodge Park as well, and she boldly supported her husband's ventures with her inheritance. Business stationery with the heading *E.D. Park Lumber Company* indicates she had a subsidiary business to Will's, and her name also appears on numerous deeds.

The couple purchased a ten thousand acre tract in Dorchester and Lyme, some twenty-six miles west of Plymouth — beautiful terrain on the divide between the Baker and Connecticut Rivers. In the midst of it lies a 160 acre lake called Cummins Pond, and nearby, the 3200 foot peak of Smarts Mountain. In 1888, the *Hanover Gazette* reported that the Park Lumber Company was putting in a three mile road near Smarts Mountain, and was planning to get its mill running shortly. The company employed 32 men, and owned 16 yokes of oxen and 16 horses.

By 1892, Elizabeth and Will had four more children: Richard, born in November of 1883, Ruth Ayer, born May 9th, 1885, Esther Marguerite, born in 1887, and Katherine, born in 1892. The happy brood flourished under the nurturing of their mother and the strong guidance of their father. Both parents filled their children with the confidence that, through education, perseverance, and faith, they could follow any pathway they chose. Anything was possible. Realizing college must be a door open to these youngsters, Will was motivated, as never before, to build Park Lumber into a business that would continue on into the future.

The time was NOW, and Park Lumber Company was on the move.

Two-Sled load of spruce, c. 1885, Warren, N.H.
with Mt. Moosilauke in the background.
*Esther Whitcher Collection*

# PART II

## An Historical Fiction of the Childhood and Education of Ruth Ayer Park

Ruth Park with her sisters and brothers circa 1895.
Back row (L to R): Richard, Mary Elizabeth, Joseph.
Front row: Katherine, Esther, and Ruth.
*Joe Park Collection*

*Chapter One*

**Childhood Affections, Plymouth, 1890**

Ruth leaned upon the arm of the upholstered bench and shifted her weight again. The high button shoes felt tight on her feet. The room was warm.

"Why did Mother make Esther and me wear these woolen jumpers?" she wondered. The tight collar scratched her neck.

"When will this be over, Mother? Esther and I want to go outside and play?" she said. Esther, her younger sister who sat in the middle of the bench, scowled in agreement.

"Just a few more minutes, children. The nice photographer has about finished. Hold still, now and look straight ahead." All six Park children groaned at the thought of having to stand still a minute longer. Baby Katherine was perched on the opposite arm of the bench, supported by Ruth's older brother, Richard. The two oldest children, Mary Elizabeth and Joe, stood like sentinels behind the younger four.

Ruth, realizing she had no choice but to wait, watched the photographer preparing his equipment. Finally, he put his head under the black cloth, called out in a muffled voice for the children to remain still for the tenth time, and then *poof* went the smoke.

Mother said, "That's it, now, children. We're finished. Grandmother will be so happy you were good. When we are in our new home in Dorchester, she will have this portrait to see you all just as you are today."

"Now, Ruth and Esther, please take Katherine and go to Grandmother while you older children come with me over to the house. More packing still to do."

Ruth and Esther ran down the dark hallway, then waited at the end for Katherine to catch up. Ruth liked the way the oriental carpets felt under her leather shoes. She stopped in the arched doorway into the parlor, where Grandmother sat in her high-backed chair, doing needlework.

"We're finished with the picture-taking, Grandma. Mother says we can go out to the yard now."

Grandmother insisted they change into play clothes, which she kept in the guest room for the children.

Once outside, leaving her two younger sisters in Grandmother's charge, Ruth ran at full speed down the lawn to the granite wall and then back again. How she loved being outdoors! It was as if she came alive — the fresh air on her face, the sounds of birds and crickets, the fragrance of flowers from the garden. Tall elm trees lined South Main Street, and the leaves sparkled in the sunlight.

Ruth knew every inch of Grandmother's yard. Hiding places abounded, behind the carriage house, and among the shrubs and trees and under the veranda on the side of the house. After hours of running and playing, Ruth especially liked to sit on the granite wall, looking at the freight wagons, coaches and carriages going by on South Main Street. Sometimes the big wooden wheels would spatter her with water. She imagined where the people were going and what was inside the tall wooden crates, crocks and barrels filling the beds of the wagons.

In the winter, the most exciting rigs were those going to Grandpa and Pa's sawmill a few miles away. These were the logging sleds, creaking under the weight of the freshly cut spruce logs. Drivers carefully maneuvered teams of oxen or horses down Main Street, past the common and the stores to the north end of town to the place where the Pemigewasset and Baker Rivers met. Here they would unload the timber in Park Lumber Company's yard.

Ruth often visited the big lumberyard with her mother or grandparents, and she never tired of seeing the men move the heavy timbers around the yard. Mother held her hand while together they would look at the big circular saws slicing the mighty trees into boards, and watch the woodsmen working the logs carefully off the rigs with their cant dogs, the long pole with the iron hook on the end. The lumber yard was alive with sounds! Clanging chains, the saw screeching, and the shouts and calls of workers, speaking in languages Mother said were French, Italian, and Swedish.

The smell, the noise, the activity — it seemed to her like the best work in the world, and she told Pa she wanted to be a lumberman when she grew up. He would tousle her hair, and pat her cheek and smile. He never told her she couldn't be just because she was a girl. That made her happy.

This afternoon, though, Ruth felt both happy and sad at the same time. Tomorrow, they were moving to a new home way out in the woods of Dorchester. When Mother told the children they were moving away from Plymouth, away from Grandmother's house, away from Pa's sawmill, Ruth wondered about these things, what they meant. What would it be like to live in the woods? Would there be other children to play with? And what about Grandmother? What would she do? Ruth was puzzled, as well, about what she heard Uncle Elbridge say: Pa was going to be a lumber king. What was a lumber king?

As her two younger sisters played on the lawn and Grandmother Dodge sat close by, Ruth lay back on the grass and watched the clouds, wondering what it would be like to be the daughter of a lumber king. Mother read to them at night in a tall story book about princesses, kings and queens, living in shiny palaces, filled with treasures. It must be like that, she concluded, only in the woods! Will we have servants to wait on us, ride in fancy carriages and have parties?

Just then, a train whistle pierced the air. It was the afternoon train roaring by on the track which lay just a few hundred feet from the street, down the steep banking towards the river. The railroad was the Boston, Concord and Montreal line and as it steamed through Plymouth, it ran parallel to the wide Pemigewasset. Ruth listened for the squealing brakes and the hiss of the steam as the engine rolled into the Plymouth station, just a short distance away. Near the big station, in a fancy two-story building with tall trees and a lawn were the offices where Grandfather Dodge had worked for many years. Ruth figured Grandfather Dodge must have been an important person because Mother showed Ruth and her sisters and brothers a cannon on the Courthouse lawn that was brought to town because of Grandfather's efforts. It had been used in a war a long time ago, and people liked to look at it to remember the brave soldiers who had fought the enemy. Ruth and the other children liked to run their hands along the smooth iron barrel and try to climb on it when no grown-ups were looking. Ruth didn't understand what war was exactly, but it must have been a serious thing for everyone's voice got quiet when they spoke of it. She had many questions about Grandpa Dodge, and she wished he hadn't died before she was born so she could ask him these questions herself.

"Names are important." She thought of how her older brother Joe was named after Grandfather Dodge. His full name was

Joseph Dodge Park. Mary Elizabeth was named after Grandmother Dodge (Mary Tewksbury Dodge) and Mother (Elizabeth Dodge Park).

One day Ruth had asked Pa, "Who am I named after?"

Pa explained. "My great-great step aunt, Ruth Richardson. And my Grandfather Park told me she was great indeed — loved everyone, especially children. Lived a long and happy life." Ruth tried to imagine this woman who liked children, was good to her family and lived long enough to become very wise.

"I want to be like her," Ruth said to herself.

The afternoon shadows stretched across the wide lawn. Grandmother Dodge was preparing to walk the three girls across the street to their home, just as she had many times before. Only this time was different. "Tomorrow," Ruth thought, "I'll be eating supper in a new place, sleeping in a new house."

Grandmother's home, the train station, the Congregational Church, and Father's lumberyard — these were the edges of her world. Ruth's brown straight hair shone in the sunshine, and her blue eyes were clear and bright. Her cheeks were flushed pink from her vigorous playtime, and her mind raced with anticipation. Ruth somehow understood that tomorrow life as she knew it would change forever.

The next morning the whole Park household was astir early. Two men from the mill arrived to load the remainder of the trunks and bundles into one of the lumberyard wagons and left for Dorchester. As Pa and Joe were hitching the team to the family carriage, Mother checked the entire house one last time. Ruth accompanied her and listened to the echo of their footsteps in the empty rooms. Without the familiar furnishings, it no longer seemed like their home. Silently she said good-bye and followed Mother to the carriage.

Ruth loved the rides in the country because of the singing. When the Parks were on an outing, Mother would teach the children songs, and the singing combined naturally with laughter. Pa said if Mother had continued her voice training at college, she would probably be another Jenny Lind. Today, once they had passed the lumberyard and were on the road to Rumney, Mother led the singing of a hymn the children had learned at Sabbath School. Pa and Joe sat up in the driver's seat. Richard, only a little younger than Joe, knelt on the seat next to Mother, looking towards his father and brother, eager to hear any tips that Pa would tell Joe about driving the team of horses. Baby Katherine cuddled in Mother's lap, and Mary, Esther and Ruth sat on the opposite seat. Ruth was thinking about the "Firsts." Today was the first time she had seen her home empty and the first time she was going to see her father's kingdom way up in the woods of Dorchester. Mother's hairstyle had come undone and curly strands of her tawny hair hung around her neck and ears. But she looks more beautiful than ever, Ruth thought, because she is so happy, her six children all around her, and her brave husband leading them on an adventure.

For many miles the road followed the gentle Baker River as it twisted along the farmland of the valley. Pa drove past Stinson Mountain, then Rattlesnake Mountain with its rocky cliffs, and then he headed the carriage west, leaving the valley floor behind. Now Ruth noticed the fields were hilly, and she imagined what fun they would be to sled on in the winter. Pa pointed out the Dorchester Town House and the village church, and still the road climbed upward. Farms and clearings grew farther and farther apart and soon Ruth didn't see any more farms. They were in the woods all the time.

"When will we get to our new home?" asked Ruth.

"It's just ahead now, dear," replied Elizabeth. "Up this next hill and around the bend and then you'll see the lake."

All the children craned their necks, wanting to be the first to see it.

Then Richard shouted, "I see it! I see the lake!" Yes, Ruth saw it, too. Through the branches she glimpsed a shimmering light, and realized this was the sunlight dancing on the water. Joe pulled the rig over to the side of the road, and the family climbed out for a closer look at the lake.

At the shore, the children touched the water, remarking on how warm it felt and asking when they could go for a swim. Then Pa showed them the dam where water flowed out and down a steep series of rocky plateaus, and disappeared in the forest. Cummins Pond was the name of this place, Pa was explaining. "We own all the land around the lake as far as you can see, up to that mountain over there," and he pointed to a wooded peak off in the distance.

"But Pa," said Ruth. "Where's the palace?" She had been looking in all directions for their new home, but she saw nothing but an old farmhouse and a barn in a clearing some distance away. Pa gestured to the farmhouse.

"There's your palace, sweetheart. Our new home." Back in the carriage, Ruth didn't say a word. When the family pulled in front of the house and began unloading, the children begged to go inside to see the rooms. Ruth joined them, but her heart sank. She thought, "This is no palace. It's just an old farmhouse, not even as beautiful as Grandmother Dodge's home on South Main Street."

The ivy-covered front porch was stacked with the trunks and furniture. Inside the front door was a large room with a fireplace, and off that a kitchen, a pantry, and a woodshed. Upstairs, Mother was telling the plan for the bedrooms. "This one is for the boys to share. This is our room, with the crib for the baby. Here's your room, Esther and Ruth. Mary Elizabeth, you now have your own room."

Ruth stood in her new room, looking out the window. She could see that shining water outside the window. Already she missed her Grandmother Dodge, and both homes in Plymouth. As she cried softly to herself, Pa entered the room and picked her up. She looked away from his face because she didn't want him to see her eyes full of tears.

"Ruth dear, when people call me a lumber king they are not talking about a storybook king that lives in a palace. They mean a lumberman who buys woodlands so there will always be plenty of trees for the mills. Mother and I bought nearly 6000 acres up here three years ago so we could increase Park Lumber Company. Grandpa Park and I have crews cutting trees way up on that mountain over there, so that we can sell the lumber to builders and paper companies. Then we will have money to do things, like send you to a nice college and buy you things you need."

He continued, "It's better than a palace anyway. In a palace you need to behave like a lady, wear your best clothes, and sit carefully on uncomfortable furniture. Here, my darling, you can wear your play clothes, and go barefoot. You can swim, and ride in the canoe, play in the woods and ride Pa's horses. Won't you try it for awhile and see if you don't like it? Even better than Grandmother Dodge's house?"

Ruth wiped the tears away on the sleeve of her blouse. For a moment, she didn't speak. Then, she said, "It's all right that we don't live in a palace, Pa. I'm not sad anymore."

She gave him a kiss on his cheek, right on the soft part above his whiskers. He put her down, and she ran to find the other children.

That night, when Pa and Mother came to say prayers with the children, and tuck them in bed, Ruth hugged them both tight. "I like our new home in the woods."

"Good, dear. So do we," said Pa. "Good night. In a few days we'll take you to see the lumber camp up on Smarts Mountain."

*Chapter Two*

**Early Lessons**

When that day came, a few days later, it was Grandpa Park who drove the children to see the camp. At first it was a trip just for the boys as Mama planned to bake blueberry pies with the girls. Ruth insisted that she must go instead to the camp, and reminded her mother of Pa's promise.

"You'd rather go there than cook with us?"

"Yes, Mama. I want to see how they cut the trees."

"Well, It's no place for a little girl. Mind that you stay close by your grandfather or brothers. Don't go wandering off by yourself."

"Yes, Mama," Ruth replied, happy that she could spend the day on an outdoor adventure.

To Ruth, Grandpa looked like Saint Nick. Short, stout, white hair and beard, but most of all he had a twinkle in his blue eyes and a laugh that made everyone around him laugh as well.

Grandpa taught the children many things about the lumber business. Names of trees, the equipment lumbermen used, and about the woods. He wanted his grandchildren to know about Park Lumber Company, and to be proud of it.

"A tote road," he was explaining, "is the road the teamsters use to bring food, tools, and equipment up to the lumber camp. Men might be up there working for many weeks, even months, so someone must bring supplies to them. Gotta have roads to haul out sled loads of logs. Stables, bunkhouses, cook shacks. Place for the blacksmith. You know children, in a lumber camp, there's a lot of work after chopping down the tree. First the branches and the top must be cut off. Then it has to be sawed into log lengths and moved through the woods. It's too heavy for men, so work horses or oxen drag it to a clearing or a yard. This is called twitching the logs out. Then when they have enough trees to fill a sled, the woodsmen load the logs with their cant-dogs. The teamsters take

# 146  LUMBER AND LOG BOOK

## Lumberman's Shanty

MANY a backwoodsman will recognize this picture of a lumberman's camp in the wilderness. No matter how poor the lumberman may be, and whatever his trials, and they are many,—whether he is known or unknown, rich or poor, in the lumber camp a stranger is made to feel at home, if worthy; if not, woe betide the weary traveller or wild woods tramp who seeks shelter beneath the hospitable roof of a chopper's dwelling.

the load to wherever the logging boss says, either the mill or the railroad station."

The three children peered into the woods as they went, testing each other on the names of the different species of trees: spruce, fir, pine, birch, maple, poplar. Ruth had already learned from her father to look for the tall straight trees for they would make the best boards. Grandpa had taught Pa, and now Pa was teaching his children to find good trees and to avoid the ones that might not be good inside, ruined by insects or blight.

Grandpa told how he and Pa would buy woodlands by going for long hikes through the woods, looking at the trees and figuring how much lumber was in the trees. Then they knew how much to pay or not to pay for the land. Sometimes it took a long time to cover so many acres and then they would have to sleep overnight in a cabin or in a tent. Ruth reminded Grandpa that he had promised to take the children with him when he cruised timber the next time.

It seemed like they had traveled far from home now. Alongside the road were blackberry bushes with clusters of shiny ripe berries. Squirrels and chipmunks skittered on the rocks and across the road. The air was fragrant with the smell of Christmas trees. So many hideouts and places to play, she thought. Pa was right. This was going to be ever so much better than any palace.

They arrived at the lumber camp, and Grandpa tied up the buggy on a rail in front of the stables. The camp was no more than a large clearing with a few low buildings made out of rough logs and boards. The structures had one or two windows, a door and an opening up on the roof.

A young woodsman came from a nearby shed and spoke to Grandpa.

"Morning Mr. Park. These must be your grandchildren. Nice to meet you. My name's Angus McMaster."

"Yes, Angus, these are three of the six. Joseph, Richard, and Ruth. Say hello, children. Angus comes from up in Canada, Nova Scotia. He's our blacksmith. Say, Angus, maybe you could show them around the camp while I go talk to Andersen and the cook?"

"Be glad to," replied Angus.

As they started their tour, Ruth knew that the first thing Joe would ask about were the horses. He loved horses, and so Angus toured them around the stables and explained his blacksmith shop duties: to shoe the horses, repair broken equipment for the teamsters and sharpen the axes each night for the choppers.

Then Angus showed them the bunkhouse where the woodsmen slept. It was cool inside, and Ruth enjoyed the change after their drive in the sun. The long room had very little natural light, just some from the doorway and a little from the smoke hole in the roof over the fireplace in the center of the room. Angus explained that the men slept on the floor, on ticking stuffed with husks and hulls and a big heavy quilt on top. The only furniture was a long table and some benches. Bags of the men's belongings hung on pegs on the walls.

On one end of the bunkhouse was a connecting room that led to the cook-shack. Here the cook stored his supplies. Angus put his hand into the large wooden barrel filled with dried beans and let them fall in a stream back into the barrel. The aroma of onions, apples, and cinnamon reminded Ruth of her mother's pantry. As Angus led the children into the cook-shack, he explained that the cook was responsible for providing three meals a day to the woodsmen.

"Those large pots, d'ya know what they be for, laddie?" Angus asked Richard.

"No, sir," replied Richard, looking at the metal cook pots with covers and long bails, hanging on the wall.

"Fer when the crew is too far away to come in to lunch. The cookie totes those pots full of hot food out to the men."

"Wouldn't they be too hot to hold?" asked Richard.

"Right," replied Angus. "So that's why they use these yokes. See, they hang a pot on each end and then put the yoke across the shoulders and carry the food without touching it, or spilling it."

The cook, a small, stocky man, with a ready grin, greeted the children, and gave them each a cup of cider and a piece of cold gingerbread. While they ate, sitting at the rough tables in the dining area, Ruth thought about what she had seen. She felt pretty important seeing another part of Grandpa and Pa's business. She could see it was a big job, keeping everything running right.

Soon, Grandpa reappeared, thanked Angus for his help, and the little group was off again, this time to inspect the work on the sledding road. The sound of axes as they hit the tree trunks echoed in the air — one, two, three chops, then again, one, two, three. Way up the hill Ruth saw a crew of three men preparing to fell a huge spruce. Other men with crowbars pried large stones out of the new roadbed while another group with shovels filled the holes in with dirt. Soon there was the call of "Timber!" and all turned toward the hill to see the mighty tree fall with a crash. A large man in a plaid shirt came over to Grandpa and tied up the horses to a sapling. The children stayed in the rig, while the men talked.

"Looks like you've got another half mile on the road done since Monday, Andersen," Ruth heard her grandfather say to the man.

"That's right, Mr. Park," replied the man named Andersen. "You've just about got your three mile addition to last year's road. Should take us real close to those spruce you want up below that ridge. We'll be done in another week. That'll give the men a chance to have a rest before the freeze comes and the real job begins."

"Good job, very good," said Pa.

Andersen continued, "I'll be sorry in a way when we leave this place. It's been a real good three years for me, and others in the camp say the same. But it'll be logged off by the end of this season."

Grandpa nodded in agreement. "Well, that's the nature of the business. We harvest the trees, and move on. We'll be needing good men to take down the mill to move it to Lincoln next summer as well as set up a camp. I know I can count on you, Andy."

"Right, boss," he replied. "I'll be there, God willing."

The two men parted and Grandpa Park spoke briefly to the road crew. When he returned to the buggy, he asked with a smile, "Who's ready for a swim in Cummins Pond?" he asked.

"Me! Me! Me!" shouted the three young Parks.

With that he swung the team around and headed down the mountain road, past the camp, and on home.

Ruth knew more than ever she wanted to be a logging boss. But whenever she told others that, except for Pa, they would laugh at her. So she kept the dream in her heart, where it would be safe until she could figure out a way for it to come true. She didn't know how, but she felt sure it would come true.

That night, after supper, Mama and Pa were relaxing on the front porch, sitting in big wicker rocking chairs. Joe and Ruth and Esther were catching fireflies out on the lawn, near the badminton net Pa had set up for them.

Ruth heard her parents talking about school. Ruth wondered where they would go to school out here in the woods. Then Mama called them all together and asked them to listen to plans for next week.

"Next Saturday, your school teacher will be arriving. She's staying with us for the fall term. The Dorchester School Board gave us permission to hire a teacher for you and the other children here in this area called District 8. Remember the school in Plymouth that Grandfather Dodge helped to establish? The place where I told you teachers learn how to teach, called the Normal School, over near the Congregational Church? Well, your teacher has just graduated from there, and so you will be her first class."

"Mama, where is she going to sleep?" asked Mary Elizabeth.

"Why, she'll share your room, dear."

Ruth asked, "But where is the school building?"

Both Pa and Mother smiled. "It's right here, in our very own house. You'll go to school right in the main room. You'll have no excuses for being late or absent, because you'll already be at school when you wake up in the morning!"

Everyone laughed. Even baby Katherine.

*Chapter Three*

**Back to Warren**

    The date was November 13, 1907, and Ruth, age 22, sat quietly in the passenger car of the afternoon train from Boston, going to Warren, New Hampshire. Esther and Katherine sat across from her. They had been travelling since morning, from Wellesley, then changing trains in Boston. Ruth looked out the window at the tree-line of the passing hills. An early blizzard had deposited nearly a foot of snow in the region, and a cold wind swept the landscape. The bare branches of the maples, the birches, the oaks, reached upward, each tree making a graceful fan-shape, silhouetted against the sky. The dull, taupe color of the trees fitted her somber mood.

    She glimpsed her reflection in the window. Her straight, dark brown hair was pinned on the top of her head. The plum color of her blouse, thin collared and ruffled in the front, reflected a rich rose hue on her image in the glass. Bright blue eyes framed with dark upward arching brows, and a matter-of-fact mouth, with a defiant set to her jaw, when her mood was such as it was today.

    As she gazed intermittently at the trees and then at her reflection, her thoughts traveled back to last year's graduation from Vassar. The yearbook committee had chosen to highlight her combative nature: "There ain't a face but what she's shook her fist in." Secretly, she was glad her classmates thought of her as a scrapper. At five foot two, it gave her self-esteem a boost, knowing she wasn't afraid of a fight.

    Katherine interrupted her thoughts.

    "Ruth, who is going to meet us at the station?" Katherine was thinner-faced, and taller than Ruth. Her chestnut hair was also in a bun. Her eyes greeted Ruth with a gentle, concerned look.

    "Pa."

    "Will he have Grandpa's sleigh?"

    "Yes."

Ruth's mother, Elizabeth Dodge Park, (1855–1907), one of the founders of the Young Ladies' Library Association in Plymouth, N.H.
*Joe Park Collection*

"Who will take Mama's coffin off the train?"

Ruth paused. Katherine was fifteen, the youngest child of the Park family, and only recently away from home for the first time, at prep school in Massachusetts.

Esther answered for Ruth.

"Angus McMaster is bringing the horses and a sled to meet the train in Plymouth. He'll take the coffin over to the funeral home, while we continue on out to the farm."

A sophomore at Wellesley College, Esther was studying nursing and already displayed the composure needed in dealing with death.

In April, their mother, Elizabeth Park, had traveled from Warren to visit Esther at Wellesley. Not only had she seen Esther, but Ruth was there as well, taking classes in literature, after graduating the previous year from Vassar College. Katherine, at the Walnut Hill School, was not far away. So, the four Park women visited frequently at the boarding house where Mrs. Park was staying. But Elizabeth had taken ill, and the physician cautioned the daughters that their mother should not be transported back to the farm but remain in Wellesley. Ruth and Esther learned what was necessary to take care of her, as instructed by the nurses, and arrangements were made for a small house to be rented for the three to stay together.

It had been a difficult routine, but neither daughter complained. Ruth cared for her mother during the week, while Esther went to classes and studied. On weekends, Esther nursed and Ruth took the train up to Warren, to check on her father, who had fallen into poor health after the closure of the Warren sawmill.

Three days earlier, on November 10, Elizabeth had died. To the Park children, it was a calamity. Their mother had been the bond that kept the Park family together. How would they carry on without her?

The gentle rocking of the train comforted Ruth, like a familiar friend. All through their childhood, the Park children had traveled on the railroad. Part of Grandpa Dodge's legacy to the Parks was a love of train travel. They went to school on the train, they went to town for errands on the train, and Park Lumber used the trains for transporting lumber to mills and wholesalers. Mama would often speak about the elegant offices where her father worked, in the two story brick building behind the Plymouth station. Ruth accompanied her there once, and remembered the ceiling and walls were made of golden oak, and the carpets were soft underfoot. From the window, she could see the station and the Pemigewasset River flowing in front of it. Out the side window was a view of the Dodge mansion on South Main Street, the stately four chimney brick home, where Mama had grown up.

The conductor's voice broke the stillness. "Ply—mouth Sta—tion." Ruth and Esther and Katherine straightened up. People they knew, who would have read in the newspaper of Mama's passing, would be getting on the train. Ruth cringed inwardly, anticipating the awkwardness of making polite conversation. Although she liked people well enough, social amenities didn't come easy to her.

Esther, everyone said, was the spitting image of Mama. Not only in looks, but in mannerisms. Both had golden curls, blue eyes, and a gracious way of making people feel at ease. Ruth relied on her to be the spokesperson of the trio.

"My condolences . . . read it in the paper . . . such a shock . . . anything we can do?" Various neighbors and friends of their parents shuffled by, tipping hats and offering a squeeze of the hand. Ruth quietly left her sisters for a few moments and walked to the end of the car. She pushed open the door and instead of going into the next car, she stepped off the train onto the platform, and looked down to the baggage car. There he was. Punctual, faithful Angus McMaster, in the driver's seat of the sled backed up to the door of the baggage car. Several men carried the huge crate containing the casket, and within moments, it was carefully loaded onto the Parks' rig.

"You're home, Mama," Ruth whispered into the twilight. She returned to her seat, as the train hissed and jolted softly out of the station. Now only twenty miles more to their home in Warren — the farm that Pa and Mama had purchased in 1900 to be near Park Lumber operations on Mount Moosilauke.

It was twilight, and the two younger sisters slept, Katherine resting her head on Esther's shoulder. *What would happen to them now?* thought Ruth. *Who would see that their schooling wouldn't be interrupted? Who would help Pa settle his business affairs?* Ruth thought of Ruth, the Moabitess, in the Bible. When confronted with the death of her husband, the woman from Moab had steadfastly refused to leave Naomi, her husband's mother. No law, custom or family pressure could sway her. She remained loyal to Naomi, following her into new places and circumstances, and met with good fortune.

*Maybe, like my Biblical namesake, I can be a steadying influence for my family in this trying time,* Ruth thought. Since graduation from Vassar last year, Ruth had been unsure what she would do. Teach

English. Study for a higher degree. Write. However, though they had not asked for it, it seemed as if Pa and Mama had needed her help. Now, she was sure she could be useful, in some way, to the family.

Questions whirled around in her mind. She thought of her two brothers in the army.

"When Joe and Richard arrive home, they will know what to do. Then the answers will come."

Ruth shut her eyes and rested her head on the soft velour fabric of the seat, and focused only on the pleasing sound and the motion of the train.

Just before nightfall, the train pulled into the Warren station. The gas lamps along the station building flickered a soft light onto the platform, and Ruth distinguished the familiar sloping shoulders and stocky shape of her father, William Richardson Park.

The three young women collected their belongings, and wrapped up in warm capes, coats and scarfs, well-knowing the cold that could await them in the open sleigh. They made their way off the passenger car to their father. Pa embraced the younger two, and then, Ruth. She glanced up to see his large face with generous features. His eyes shone with tears.

"Oh my girls, how can this be happening? It's like a terrible dream that I can't wake up from."

Not knowing what to say, they just hugged him tighter. As if demanding an answer to an unanswerable question, he continued, "What are we going to do without Elizabeth?"

Even when his own parents had passed away, Ruth didn't remember seeing her father cry.

"Let's get home, Pa," said Ruth, after a few more silent moments. Whereas the two younger sisters were better able to speak with the neighbors on the train in Plymouth, Ruth had always been the one to take charge in matters dealing with Pa. He was strict, at times, and the children treaded softly around him. But Ruth had learned from an early age to stand up to him, and not act afraid, and he responded positively to her forthrightness.

Ruth began shepherding the family toward the family sleigh. As the others loaded in, Ruth climbed into the driver's seat, alongside Pa. "Let me drive, okay?" she said quietly. He handed her the reins and moved over. She was glad to have something she could do that would be useful and demand her attention.

"Get along there, Molly. Go on now, Dorcas," she called out to the horses. She directed the rig out into the main street, passing within minutes in front of Pa's sawmill. The gold letters shone faintly. Park Lumber Company. The building was dark. The fading rays of the sun shone across the snow covering the empty millpond, where a million feet of spruce logs had once floated. The tall smoke-stack rose in the evening sky like the mast of an abandoned ship.

The horses pulled the sleigh easily now over the packed snow on the road, and the new snowfall muffled the sound of the hooves. On they went, past the company store, the boarding house, the large stable that had been home to a hundred head of horses. Past the pulp mill — all dark. Ruth swallowed hard, and kept her eyes focused on the road ahead.

She knew there would be signs of life at the farm, and that was where they all needed to be. Her sisters seated on the back seat, talked softly. Pa stared straight ahead.

As the sleigh came over the top of the rise, northwest of Warren Village, Ruth slowed the team. It was always her moment to take in the view. Off to the northeast, glowing pink in the last tones of the sun, stood Mount Moosilauke. Directly ahead lay Park Flat, acres of level farmland that her parents had purchased in 1900. Off to the west, hidden in the dark trees, flowed Ore Hill Brook. In the middle of the flat were the lights from the Park Lower Farm, where the hired help lived, and further on, at the end of the flat, more lights pierced the growing darkness. The Upper Farm. Home, at last!

They traversed the quiet valley, and when the sleigh had arrived at their own front porch, Esther, Katherine and Pa alighted and unloaded the luggage.

"I'll take care of the horses, Pa," said Ruth. "Go ahead. I'll be in soon."

Even though she wasn't dressed for barn work, Ruth was relieved to be alone after a trying day.

She unhitched the sleigh in the shed, and pushed open the large sliding barn door. Her cold hands fumbled with the matches as she lit the old lantern. It made a scratchy sound as she hung it on the rusty hook on the wall. She threw blankets over the horses, filled their buckets with water from the pump, and pitched them some fresh hay.

"Good night, friends," she said, as she blew out the lantern. She closed the door and headed for the house. On the porch, her fingers resting on the door handle, she paused, took a deep breath, and went inside.

*Chapter Four*

## The Family Gathers

The house was warm, and steamy with the smell of food. A neighbor, Mrs. Shortt, who helped with the cooking and cleaning when Ruth was away, was baking a ham for dinner. Ruth could hear her pleasant voice in the kitchen, and she also heard Katherine talking, and then the very deep voice of her older brother, Joseph Dodge Park.

Joe was a medium built man with dark, straight hair, and a remarkably bright, wide smile. After a few moments of affectionate hugs and teasing, Ruth, Joe, and Katherine sat in the dining room to talk, before the meal. Pa and Esther had gone upstairs to rest. Joe continued to quiz Ruth, playfully. "You look so pretty, young lady. You must have a special fella?"

"Don't have a fella, Joe. I'm waiting for you to introduce me to one of those soldier buddies of yours."

"Matter of fact, I do have someone in mind. Not a military man, though." He winked at her, and grinned.

The conversation turned to the family matters at hand, the loss of their mother and the arrangements for the service, and Pa's condition.

"Ever since the mill closed," Katherine explained, "Pa's been out of sorts. He blames himself, says people will think of him as a failure. He's sure we'll all have to go to the poor house. No matter what you say, he won't change his mind."

Ruth nodded her head in agreement. "And now, with Mama gone, he's lost the hope that he could make it up to her."

From the start, Mama and Pa's courtship had caused a stir. He, the handsome son of William Richardson Park, Sr., the timber manufacturer from Haverhill, who had recently moved to Plymouth and she, the only daughter of the B. C. & M. general manager. When the Dodges' only son, John, passed on in 1877, which made Elizabeth the primary heir, Joseph and Mary counseled her

Ruth's brother, Joseph Dodge Park, West Point, 1904.

on the wisdom of involvement with the lumberman. It was a foreign world to the citified Dodges, who perceived the lumber trade as rough, risky, one that attracted unscrupulous characters. Wouldn't she be better off marrying a Plymouth man, someone well-known to the family, perhaps, a doctor or attorney?

But the more they attempted to influence her against him, the closer Will and Elizabeth became, and in June of 1879, they married in the Plymouth Congregational Church.

Will's success at Brown University, and his awareness of the trend toward large-scale industries, motivated him to expand Park Lumber operations. The company branched out from Plymouth

Richard Park, West Point, 1907.

into Rumney, Dorchester, Lyme, and Lincoln.

Never afraid of hard work, he blasted into a routine of long days and full weeks, vigorously taking on more and more of the management of the company. By 1900, both his parents had passed on, and he became sole owner. Two others had also died that year: Mary Tewksbury Dodge, Elizabeth's mother (Ruth's grandmother) and Mary Elizabeth Park, Will and Elizabeth's eldest child (Ruth's sister), succumbing to an epidemic.

Will believed it would only be a matter of time before he could give his growing tribe the same lifestyle Elizabeth had enjoyed as a girl. But to succeed, he must maintain the vigorous pace. Lumbermen had to go where the tall spruce trees grew, and rather than be left to raise her children essentially by herself, Elizabeth chose to go with him. So the Park family moved out of Plymouth to be near the lumbering operations. First to Dorchester, and then, in 1900, to Warren.

Elizabeth brought beauty and refinement into the children's lives, despite the lack of elegance in their rugged surroundings. She cultivated within them an appreciation for literature and art. Not only could she sing beautifully, but compose as well, taking the words of poems written by one of her friends and setting them to music.

By the turn of the century, the Parks were well on their way to completing the transformation from the lumber mill of the past to the mill of the future. The supreme test came with the purchase of three thousand acres on Mount Moosilauke, a 4800 foot peak of the southwestern White Mountains.

Park purchased a burned-out mill in Warren on Black Brook, close to the railroad station. Two huge boilers were bricked in for power, making it the largest steam operation in the region. Four railroad sidings ran alongside, and the millpond and dam were improved. Three lumber camps were established on Mount Moosilauke. A slasher mill was built to take the sixteen foot logs out of the giant spruce, and saw the tops into four-foot logs for pulp. The logging animals were kept in an immense stable Park had built in town, not far from the company store. It was a total commitment to a venture, an industry and a town. To cut the costs of transporting the logs from the three lumber camps on the mountain, the Parks built a 3 mile light-gauge gravity railroad down one side of the mountain.

But despite their best efforts, several factors conjoined to prevent success: the general waning of the timber industry, and the national economy, and too fast an expansion without sufficient adjustments to the internal structure of the company's accounting methods.

By May of 1904, an attorney from Laconia, W. D. Veazey, and his accountant, Simeon Frye, were assigned to take charge of the Parks' books in order to protect the creditors, until debts were satisfied. The mill closed, and the Park holdings became encumbered with mortgages. Elizabeth, once one of the wealthiest women in Plymouth, found herself asking for credit from merchants.

However, the children only heard about the problems, as they were away at school: Joe and Richard, at West Point; Ruth, at

Vassar College; Esther, at Wellesley College; and Katherine, at Walnut Hill School.

"What about the logging jobs? And the farming?" Joe asked.

Ruth responded, "The farm is lying waste. The logging boss, Angus McMaster, from Nova Scotia, and the teamster, Alfred Dale, and a few local hands are doing whatever is being done. With Angus's direction, they manage to keep things going.

"One immediate problem is the horses. We've got twenty, still, left from the Moosilauke logging, and there's no market for them. Workhorses don't bring anything at the auctions, and how are we going to keep feeding them? Can't abide the idea of destroying them.

"The only logging going on is one lumber camp drawing pulp wood over in East Warren. I've been out with Pa to check it. Angus says it's about finished. Angus told Pa we'd probably clear enough to pay the help."

"And the real estate?"

Ruth sighed. "It's all a jumble. Mortgages, back taxes. Bankers and attorneys seem to be firing at us from all angles. The attorneys that were supposed to be helping us have never given us a clear picture of where things stand. What really bothers me is one firm bought some of the land from Pa and Mama when this happened, the assigneeship, and they're now doing some high-tone logging operations themselves, on land they really stole from us."

Ruth's voice, usually high pitched, was getting even higher pitched. Joe put his arm around her. "Take it easy, sis. We'll figure out something. We will. Let's get the arrangements for the service taken care of, and then we can talk to the attorneys and see what we have to do. Richard and I may have to help out with our salaries for awhile. It's going to be all right."

He embraced his sisters again. Katherine went upstairs to call Pa and Esther to eat. Joe began carving the ham.

"We'll circle the wagons, Ruthie. You'll see."

"I hope you're right. At least now, we will have all the family home to make some decisions. I'm really glad you're here, Joe," she said.

The table was set with Mother's best china. The candles in the centerpiece cast a soft light on the walls of the dining room. There hung the paintings Mother had done. Ruth brought the food to the table, and looked around the room where the family had gathered so many times before. A glimmer of hope ignited within her.

"In some manner, albeit changed, we will continue to gather as a family. As long as we have a family, we have not failed."

# 8  LUMBER AND LOG BOOK.

**Logging in Winter.**

MANY a pleasant day, as well as one of toil and labor, have lumbermen spent at such a place as this.

*Chapter Five*

**Timber Pirates**

Ruth woke at dawn to discover that it had snowed again during the night. Fence posts, well covers, the barn and silo roofs were freshly covered. No wind stirred; all was quiet.

Without waking anyone, Ruth went out on the porch and put on her snowshoes. Across the lawn, and the road. From there, she traversed the fields and moved quickly into the woods towards Ore Hill Brook. Following an ancient logging road, whose pristine snow-covering was broken only by the tracks of a fox and smaller creatures, she heard the muffled conversation of the brook, which was partially frozen, with white plateaus of ice jutting out from the rocks. The water ran on through the forest, like a black shiny ribbon. Crossing the brook where three small logs had been laid as a bridge, she headed up the ridge.

Within the hour, she had gained her destination, a clearing where the whole valley called Park Flat was visible. This spot had served for a retreat on many occasions, a place for solitude and reflection.

Smoke from their farmhouse chimney below, swirled slowly skyward, as she had stoked the fire in the wood stove in the kitchen before she left. In the middle of the flat, the silos, sheds and house of the Lower Farm clustered together in the broad expanse of white. Off the house protruded a long ell called the Men's Quarters, and this was where Angus McMaster lived. Ruth knew he would soon be riding up to the Upper Farm to care for the animals. In all the uncertainty of the previous year, Angus had been the one constant. A disciplined, hard-working man, rising before dawn each day, fulfilling his duties in a methodical way, he never interfered with the Parks' business or offered unsolicited advice. But he was aware of all that went on, so when a need arose, he could respond without belabored instruction.

A Scot from Nova Scotia, who loved the lumber woods, he had

followed the work down from Canada, to Maine and then New Hampshire. He was not yet twenty years old when Grandpa Park had hired him for the Smarts Mountain job. Quickly, he learned blacksmithing, and to be a woodsman, and then woods boss, taking charge of the large camps on Mount Moosilauke, in the early 1900s.

When the mill and camps closed down, Pa had no choice but to lay off scores of workers. Angus left to find work in the White Mountains. He handled large crews of lumberjacks, but when business fell off after 1907, Angus returned to Warren and looked up Mr. Park.

"Got no way to pay extra help now, Angus. Much as I could use your help, couldn't do it," said W.R. "You better keep looking."

"Where would I go?" he'd said to Pa. "No, Mr. Park, I'm good enough with an axe, and a fair blacksmith. You give me a place to stay through the winter, pay me when and what you can and I'll tend to your horses, and whatever logging you have to do."

And that was the arrangement that continued now. Angus bunked in the Men's Quarters, conferred with Pa daily, and kept things running.

Ruth had known Angus since the early days in Dorchester, and he always looked out for her, like a kid sister. Now, as a young woman, she realized he was only fifteen years her senior. A broad-shouldered, muscular man, with dark hair and blue eyes, he stood a head taller than Ruth's five foot two inches.

It was Angus's voice that, even as a young girl, had arrested her attention. Of Scottish heritage herself, her great-great-grandfather, William Park, having come over from Glasgow, her heart thrilled to the pleasant rhythm of the Scottish speech. The musical variations of high and low tones, and elongated vowels and distinct consonants intrigued her.

Angus was a confident, animated person. Few could surpass him at telling a story. With untutored skill, he knew to keep his face straight and unsmiling, as he delivered the punch-line. And then, after his audience had exploded with laughter, he would laugh as well at the absurdity of his droll tale.

Like many woodsmen from away, he liked to talk about life back home. "I'm the youngest of nine boys and one sister. I'm me mother's favorite, her baby. There's another Angus, too, me oldest brother. We'd sort of run outa' names when I'd come along so my dear old mother just started in again. But I was always her favorite. I remember the times I've held the yarn while she knitted socks for the whole family."

He had nothing to hide. He wasn't afraid to be himself. Ruth admired this quality that would allow a hard-working, capable man like this to freely admit his tender feelings for his mother.

Not that he wouldn't glower and fume if something wasn't going right, like work falling off or misbehavior. Once, while Pa had given the crew the weekend off, a few of them started acting up at Samaha's Store in Plymouth. Realizing they were drunk and argumentative, and threatening to do each other harm, the young nephew running the store had sent for Angus McMaster, who was also in town. Angus arrived, well-dressed, and sober, and brought the situation under control. After paying young Samaha for any trouble, he took the men back to the farm. In the morning, he scolded them heartily, and collected back the monies owed.

Ruth began her descent toward the valley, but then decided to go a little further up the ridge to view the western slope, where the Parks owned a 100 acre tract. She'd overheard Pa telling Angus that this would be the next lot for their crew to get the trees down. She snowshoed quickly for half a mile. At first, when she heard the sound, she thought she must be mistaken. She stopped. There it was again — the familiar echo of someone chopping a tree with an axe. This was the Parks' land, and no one had given orders to begin logging here. Who could be out here?

Suddenly, all her senses were alert, and she picked up her pace. The sound grew louder. Within minutes, Ruth's fear was confirmed as she saw where a lumber crew had been at work. To the west of the trail lay a newly created yarding area with a tall stack of spruce logs. In every direction were boot-prints, and the paths of logging sleds being dragged by oxen.

The sound of chopping had stopped, but to the east she could hear a crosscut saw and men's voices. She hurried toward the sound and saw a crew of three woodsmen. One held an axe and stood by a felled spruce. The other two sawed. Two yoked oxen stood nearby beside a pair of sleds ready to be loaded. Clouds of steam rose off the animals' backs in the cold morning air. A large warming fire of mostly hardwood logs snapped and crackled.

*Compose yourself, compose yourself,* rang a voice within her.

"What in the blazes do you think you are doing here?" she cried out, still at a distance but quickly advancing.

The crackling fire and the crosscut saw had apparently drowned out her breathy, high-pitched voice, and the men took no notice. Drawing all the air she could into her lungs, she bellowed the question again.

She was now only twenty yards away, and this time the men heard her. They turned toward her, momentarily stunned to be accosted, and by a young woman at that.

The head chopper swung his axe hard into the tree, intending to rest it there, and walked from the center of activity towards Ruth. He was a barrel-shaped man, with a long black beard that fanned out around his neck and upper chest. He stood a head taller than her. He greeted her with a grin and seemed amused at her turmoil.

"Why, Miss Whatever Your Name Is, don't come hollering around my crew like that. You're liable to spook the oxen and cause us a heap of trouble."

Anyone knew that oxen aren't easily spooked, and she recoiled at his condescending manner.

"It's Miss Park, and you don't know the meaning of the word trouble unless you can tell me what in the blazes you're doing logging on my father's land!"

He reached for his shirt pocket, and carefully unbuttoned it. Ruth noticed he was missing his index finger on his right hand. He brought out a folded paper, and opened it up. "We're logging this land by the authority of the landowner, and I have a signed contract to prove it."

Ruth's hand trembled as she took the paper. She knew without a doubt this was Pa's land. Throughout the assigneeship, many woodlands had been mortgaged, but the payments to the banks had been kept current. Every dollar from the recent logging efforts had gone solely to pay the bank and taxes on the various tracts. The Parks had not even proper wages to pay Angus and Alfred, and the family suffered for normal groceries and supplies, but the monthly notes had been met.

The contract looked false to her from the minute she saw it. The lot number had been scratched out, and written over. The landmarks describing the property didn't fit anything in the vicinity. Ruth had walked these acres with Pa and Grandpa Park enough times to know there was no similarity between what the paper was describing and the land they were standing upon. It began to dawn on her that these were timber thieves, illegally logging her family's land.

"This is our land. Park Lumber owns this lot and my father never gave anyone permission to log here! You and your men had better get out of here right now and quit stealing our trees!"

The other woodsmen listened with interest at the exchange of words, but said nothing. Their boss was no longer grinning, but instead, growing flushed in the face.

"Young lady, I don't know where you get your information, but I have orders from the landowner to log here, and I'm losing money while we debate this, and I'm not going to debate it any longer. So, if you'll pardon us, we're going to get back to work."

Ruth's eyes flashed about the area. She must stop them from taking any more timber from the Park land. Each log they removed represented dollars the family sorely needed.

Ahead of her, resting against a tree trunk, was a rifle. Woodsmen often kept one handy for shooting deer.

Quicker than the chopper because of her size, Ruth raced ahead of him. She grabbed the weapon and jammed the butt-plate against her shoulder and took aim at the man. For the second time, the men stood stunned.

"Throw the sleds on the fire!" she shouted.

No one moved. The chopper grinned again. "You can't do this. Give me that gun before you hurt someone."

Ruth held firm. "And don't think I can't shoot this thing, because I can! Get both those sleds on this fire NOW!" She kept the gun barrel pointed at the chopper's belly and beckoned toward the fire with a quick jerk of her head.

"I'm not fooling with you, Mister!"

The chopper nodded to the other two men to do as she demanded. Together the woodsmen picked up one of the heavy sleds – about three and a half feet long and five feet wide, its runners sheathed with steel — and heaved it onto the fire. There was a tremendous shower of sparks. Then they threw the other one onto the rising flames. Without the sleds, the thieves were out of business.

"Now take your oxen and get out of here!" She circled slowly around the men, keeping the gun leveled at them. The two woodsmen put on their packs and picked up their saw and axes. The head chopper stood in place, fuming at the outcome of events, and not ready to admit defeat.

"Come on, bud," one of the other men said. "She's just crazy enough to pull that trigger. Let's get out of here."

Reluctantly, the leader grabbed his axe out of the tree and prepared to follow his two men, who had already started goading the oxen down the slope.

"I'll have the sheriff after you!" he shouted. "Stupid, idiot girl!"

Ruth watched them descend about halfway across the cleared slope. When they slowed and seemed to be fishing about for something in their packs, panic struck her. Was it another weapon? All along, she had felt certain she could outrun them, if they pursued her, because the snow was deep, and they were in boots, and she was quick on snowshoes. But, now she knew she needed to be out of range.

Without another moment's deliberation, she turned and hurried uphill, retracing her tracks. Once out of the clearing, she heaved the rifle into the brush and began running home.

*Chapter Six*

**Safe at Home**

Ruth was covered with perspiration when she returned to the Upper Farm. She quickly took off her snowshoes, and ran into the house to find her father.

"Pa! Pa! They are stealing our timber!" she shouted.

"Who is? Where?"

"Over on the west side of the ridge, going up to Ore Hill. Three men with two oxen. They had a big pile of spruce there already, but I grabbed their rifle and made them throw their sleds onto the fire, and clear out. Then I ran back here to get help. You've got to stop them, Pa."

"Sounds like you already have, Ruth," said Pa. "Go tell Joe. I'm going after Angus and Alfred. You stay here." As he left the house, he called back, "Richard is home."

Ruth ran to the kitchen, and blurted out the news to her siblings. Immediately, Joe and Richard put on their jackets. Richard gave Ruth a quick embrace as he left the house with Joe.

Within a few minutes, the men left the yard in the Parks' large sleigh, pulled by two teams of horses.

Ruth sat down by the wood stove, perspiring and exhausted. Her sisters helped her peel off the wet clothing. They brought her a basin of warm water, a blanket and a towel. Ruth washed the sticky perspiration off her upper body, and wrapped her wet hair in the towel, and pulled the blanket around her. Katherine handed her a cup of hot tea.

"Weren't you afraid?" asked Katherine, after Ruth had told them the account in more detail.

"I was so mad I didn't have time to know if I was afraid. When I heard them cutting down our trees, and I knew so well all that Pa and Mama had done to hold onto what land they could after the assigneeship, the idea of someone stealing the timber off it was more than I could bear."

Upper Park Farm, Warren, N.H., burned April 13, 1913.
*Esther Whitcher Collection*

An interior of one of the Park households circa 1910.
*Park family collection*

Esther gently chastised Ruth. "You took a lot for granted. You assumed that you were right and they were wrong. You assumed the rifle was loaded. You assumed you could outrun them. You could have gotten hurt, Ruth."

Ruth nodded her head. "I know."

Pa and the other men returned several hours later. The timber thieves were nowhere to be seen. Angus told Ruth the large stack of long logs stood as she had seen it, and all that remained of the sleds was the ironwork and a little charred wood.

"They weren't going anywhere on those sticks of charcoal!" he said, with a laugh. "Aye lass, it took a stout heart, what you did there, out of love for your Pa. I don't think he knows yet whether to take you over his knee for scaring him half to death or to hug you for saving the day!"

Pa and Ruth made plans to report the incident to the sheriff in Haverhill the next day, complete with Ruth's descriptions of the men, so that the proper authorities would be aware of the illegal activity. New boundary signs would be made and Angus would post them, stating that the lands were owned by Park Lumber Company and that there was to be no trespassing.

By supper that night, things had returned to normal, and Ruth had time to converse with Richard. He was a stocky, fair-haired man with a pleasant face and a noble brow. He and Ruth, being only eighteen months apart in age, had been close friends as youngsters, and over the years, had shared many interests, including the outdoors, sports, music, and literature. Now, he told her, he was enjoying his army life, working as a civil engineer. He expressed regret that he couldn't be more help at home, but that he was absolutely certain that the army was his calling.

He had recently wedded a woman from Portland, Oregon, named Winifred Higgins, and they were very happy. Together, they were making friends both in the military and in social circles in the Portland area. He said they were hoping to start a family soon.

Before bed, Ruth wrote in her journal: "I want nothing but happiness for Richard and Win. He has a new life, and experiences that are far away from when we were growing up in the lumber woods and on the farm. Maybe, he will change his mind, once the newness of his career has worn off, and come back to New Hampshire. But then, maybe I must admit the truth that I am the one who loves this farming and logging life. I am never happier than when out of doors, and with hard work and purpose ahead of me."

Mt. Moosilauke from the Park Farm.
Warren, N.H.

*Chapter Seven*

**Logging Days**

   After Elizabeth's funeral, the children had begun the process of putting their lives back together. Everything felt so different without Mama. The house seemed too still. Where was the rustle of her skirts, her cheerful greetings to all in the morning, the familiar humming as she worked in the kitchen? In their own ways, the siblings tried to fill in the void: Esther found she possessed organizational skills, and coordinated the reception after the service; Katherine assumed the role of cook, planning the menus and shopping; Ruth, while housecleaning, located the songbooks of old ballads and restored the tradition of family sings after dinner.
   Joe was the executor of the insolvent estate, an estate with more than seventy creditors, ranging from a few dollars to many hundreds of dollars. With grace and tact, Joe visited each of these parties, local business owners, people who had always admired Mr. and Mrs. Park, and explained that the family was working to liquidate assets to pay the debts.
   The brothers were concerned. How would Esther and Katherine finish their schooling? Knowing their mother and father's strong faith in education, Richard and Joe sacrificed to pay the tuition and expenses from their army salaries.
   Ruth resolved that Pa would benefit from less time by himself and more time with others. Each day, she would think of ways to bring him into the center of the activities. She encouraged him to accompany her as she transported family members to the train station, as they departed from Warren to return to work and school; Richard, to Fort Flagler in Washington, Esther and Katherine to school in Massachusetts. Even Joe had to leave for a few days, to make arrangements with the Army at the Quartermasters Station in Boston for an extended leave.
   Angus had taken a crew to bring out the logs the timber

thieves had yarded. There was a considerable amount of spruce in the area that could be cut as well, a job that would last several weeks, so Angus rounded up a crew to set up a camp, consisting of a bunkhouse, a cook shack, and a stable. Pa had planned to help them, but his strength wavered on the day they left, and he stayed behind. Now, feeling stronger, when Ruth suggested they take a drive up to see how the work was going, he agreed to go. On foot, Ruth could reach the area in an hour, by going over the ridge, but going by roads meant doubling way around the back of the ridge, and it took twice as long. Mid-morning, they arrived in the camp on an old logging road.

Angus was working near the road, some 50 yards away, walking alongside a team of horses in the process of hauling a log up an embankment. He waved a quick hello, careful not to take his eyes off the situation for long. One couldn't be too watchful when around these unpredictable logs and sometimes equally unpredictable horses. "Back now, back, WHOA," Angus continued his conversation with the horses who worked on a combination of voice and rein commands.

Meanwhile, Ruth walked to where the other lumberjacks were working, and saw a newly hired man, actually a young boy, logging with another pair of horses. He struggled with a log stuck in more than a foot of snow and mud. The chain repeatedly slipped off each time the tension increased. In a few moments, Ruth was able to show him how to wrap the chain in the way Pa had shown her.

On the next try with the horses, the log didn't slip, but slid easily behind the team. When the young man had safely moved the timber to the yarding area, he smiled and said to Ruth, "Thanks. I didn't know girls knew how to log."

"Some do," she replied.

This surprise at her knowledge of the logging world was not uncommon. Ruth had encountered it frequently in lumber camps and sawmills. Sometimes the comments were friendly, as this young man's; other times they were edged with hostility. More and more, Ruth was thinking about her future. She was learning that she loved the out-of-doors, that she had a head for numbers, and she thrived on hard work. She longed to find her own path in solving some of these questions, and as soon as Joe and Richard settled the estate, then she would get on with her life. For now, she was useful, and that was enough.

As they were preparing to return home to Warren, Ruth said to Pa. "Let's go see if we can find that rifle I pitched into the brush."

Within a few moments of pushing away branches and weeds, they located it lying on the snow. Ruth picked it up. "Winchester Model 1886," said Pa, an expert marksman, who had taught all his children to shoot. "It's a popular model."

"Look, Pa!" Ruth exclaimed as she turned the rifle over. Carved in the stock was the name "Adam Evans." Five distinct notches followed the name. The last notch was fresh, whereas the preceding four had darkened.

"Well, I'll be!" said Pa.

"What are the notches for, do you suppose?"

"Could be a hunting record. Maybe the number of deer, or bear he'd shot. Wonder why they didn't come back and get the gun."

"Maybe they didn't see me throw it away. They were a good distance down the slope by then."

"Could be. At any rate, it's very good evidence and we'd better get it to the sheriff right away. If we can help his department find these robbers, we'll be doing the landowners and the loggers in the region a favor."

The next day, Ruth and Pa traveled to the Grafton county sheriff's office in Haverhill, to turn in the rifle as evidence to corroborate their report of the week earlier regarding the theft. They were told that Adam Evans was a notorious timber thief and that authorities in both New Hampshire and Vermont were pursuing him.

When Joe returned to Warren from Boston, Ruth and Pa told him all the news of the lumber camp and the discovery of the rifle. He gave them a big smile, and said, with a wink, "I can see it now! We'll hang out a shingle: Park and Park, Detectives for the Lumber Trade."

Park Lower Farm, 1918, Warren, N.H.
*Photo courtesy Rhoda Shaw Clark*

*Chapter Eight*

**New Beginnings**

Joe put in duty time at the Quartermasters Depot in Boston, and took the train back to Warren to work on finalizing Mrs. Park's estate. After a year of this routine, he was growing anxious. It was obvious that he didn't take to the farm life. At night, he paced the floor of the kitchen, like a panther in a cage.

"This life's not for me, sis. My place is with my unit. That's where I belong. I love the army, the discipline, the challenge. That's my life. This is Pa's dream, don't you see? This farming, logging life, it's not mine. You understand, don't you, Ruth?"

Ruth stared at him without answering. He went on, "For me, there's no glory in accomplishing things done by others. I want to do the unique, the unusual, the extraordinary. I feel those opportunities are waiting for me in the army, not here."

In September of 1909, while he was in Warren, Joe's superiors notified him that he was to be transferred into the Fourth Cavalry. As a boy, he had spent every free moment in the stables. He watched the hired hands handle the work horses; he'd mercilessly pestered the teamsters with questions about driving team. If one of the Park children was given the stall-cleaning chores, and needed help, Joe volunteered. Saddles, tack room, the barn — this was Joe's world. His greatest thrill was riding at full gallop across the meadows and on the country roads.

This news from the Army revived his spirits, and Ruth realized, with great regret, that Joe was not going to take over the Park affairs.

Before he left Warren for his new assignment in the Philippines, he appointed Richard and Ruth co-executors, even though Richard was far away from New Hampshire.

Joe consoled Ruth. "You can do it, sis. You like the woods and the farm. Pa listens to you, and if you just keep at it, some of the land will have to sell. Then, you can settle this estate up in no time, and get out of this valley. See the world, go where you want, and start a life of your own."

Esther and Katherine returned to school, and now Joe was gone. Ruth and Pa were the only residents at the Upper Farm. At night, when Pa was in bed, Ruth would sit at the kitchen table. She kept all the grocery advertisement sheets, which were blank on the backside, and she would list all the debts and assets of the estate. The outgo was greater than the income. Each month that passed put more of the family lands in jeopardy of foreclosure or tax sale.

This quarter, mortgage payments and taxes owed were in brackets many times over what she and her father were able to realize from their minor logging efforts. What they could pull together only sufficed to pay the help, and buy feed for the horses and groceries for the household. Wood lots in the nearby towns, the large tract on Mount Moosilauke, and the extensive acreage at Cummins Pond in Dorchester, where the original family homestead was located, were devalued greatly by the fall-off of the lumber business nationwide. Most of these acres had been stripped of the spruce trees during the early years of Park Lumber Company, when Pa and Grandpa had been working with portable mills and large crews.

The age of the timber baron was over. Lumber prices had reached a peak in 1907, and had been steadily dropping since. Many investors, lumber companies, and landowners were left with large holdings of woodlands with no way to pay the taxes. In many cases, if back taxes were not paid, lots would be forfeited to the town. For the Parks, the bank holding the greatest number of mortgages was the Peoples Guarantee Bank in Laconia, and the president was a man named Mr. Simeon V. Lyeman. Ruth traveled by train to meet him, and to sign estate papers.

Lyeman stood over six and a half feet tall, with hands nearly the size of tennis rackets. He was dressed in a dark suit, a silk vest with a gold pocket watch, starched shirt and bow tie, and glistening leather shoes. The paperwork taken care of, he looked over his glasses at Ruth and said,

"You and your father are going to have to sell these lands at a loss. You don't have any choice. Whether you know it or not, young lady, you're in big trouble, and it's only going to get worse. Nobody wants these lumber lots anymore. You're in the wrong business at the wrong time. The best advice I can give you is to sell at whatever price you can, and get out of this logging life. It's no place for a woman."

His cold, stern manner caught Ruth off guard. She was not expecting any judgement, opinion, or evaluation of the Park finances, only to sign some papers. She felt humiliated, not so much by what he said, but by her inability to find the right words to answer him. The appointment left her feeling inadequate, and insignificant.

On the way home on the train, she thought, "How can he talk about our situation like that? Like we were paupers? When Park Lumber had a million board feet in the millpond, you would have been singing a different tune, Mr. Lyeman. You'd have been wanting Pa to bring business to you, along with all the accounts and payroll and company store receipts."

Lyeman's words haunted her that night, as she sat there at the table, pencil in hand. Sancho, the big, black Newfoundland dog, nestled his head in her lap as she worked on the accounting.

"What do you think, Sancho? Do we have a choice?" She petted him, tenderly. Sancho looked up at her, and whined. Ruth said softly, "There must be a way, big dog."

She propped her elbows on the table edge, put her head in her hands, and closed her eyes. Certainly, with all the intelligence in the universe, some shred of it could be available to her and her family at this time of need. For many minutes, she remained still, her spirit aching for answers.

Then, as if a voice had spoken it, she heard, "Speak to the earth and it shall teach thee." She looked up, arrested by this phrase from Proverbs that she had remembered from her college Bible class. "Speak to the earth and it shall teach thee." Several times, she repeated the statement.

"How can the earth, the soil, the ground, teach me?" Out of the window, she glimpsed the moonlight on the fields of Park Flat, the trees of the woods backing up the ridge and far beyond the slopes of Mount Moosilauke.

"What am I not seeing? Income from lumbering, peeling poplar, pulpwood, cordwood." She went over those figures again. A shimmering idea danced. It seemed so frail, what was it? Fields of crops. Like in the old days, when Grandpa Park and Pa had produced crops to feed the family, the crews, and the animals.

Ruth said to herself, "Crops could bring in some income. It would reduce the cost we pay now for the animals, and if we produced more than what we need for ourselves, we could sell the rest."

Suddenly, the notion of raising crops brought another idea: leasing out the teams of horses. Certainly, business people in the region often leased horses for short-term ventures. In the lumber world, the wholesalers from the big cities were always coming by the logging jobs, driving a team leased from the livery in town. "Why not OUR horses?" she thought.

Now, a whole festival of ideas filled Ruth's mind. Producing corn, oats, potatoes, cabbages. A dairy herd. Selling the extra milk to the creamery. Ruth envisioned the farm in full production.

That's what Grandpa used to do in the years after the Civil War, grow all he could on the farm in the spring and summer. Winter was time for the lumber woods. All his labor in the growing seasons, creating feed for the horses and food for the workers, supported his production of lumber.

"I could start small, a few chickens and cows. By selling the excess milk to the creamery and eggs at the market, we could help pay the household expenses, while making more on the woods work. What about the old equipment, the harrow and hay rakes? Maybe some of it can be revived. With the costs reduced for feeding men and animals, we could then afford to hire more men to get the trees down, which would generate money to pay the loans, and the taxes."

For the first time, she saw a clear path to restoring life to her parents' dream. Her pencil flew over the paper, as she pinned the idea down, making a list into practical steps, and what each step might yield in income, and what it would cost. Flushed with new hope, she thought of waking her father to impart her enthusiasm.

But past experiences had taught her to keep the idea to herself, to keep warming it with facts and figures until the time was right. Like a college paper, she needed to do careful research and then make an intelligent presentation.

Before going up to bed, she went out on the porch and looked across the valley, drinking in the fresh night air. Stars brilliantly shone overhead.

"It would work. I know it would. God help me to accomplish it."

The next day she wrote to the University of New Hampshire for information on the new methods of farming, animal husbandry and raising poultry. A large packet of catalogues and booklets arrived a week later, and at night, she studied the material. Then, she wrote her master plan, debts on one side, estimated profits on the other. For the first time, the one side equaled the other.

"Not a profit, but it's a plan to get us all fed, with enough to keep the taxman and the bankers happy until some real estate sells. Now is the time to present this to Pa."

She also decided to reveal the plan to Angus McMaster at the same time. Pa listened to Angus, and respected his judgement, so in the morning, she invited Angus to come to the Upper Farm for supper. Mrs. Shortt agreed to make an extra special meal, including biscuits, savory roast chicken, mashed potatoes and gravy, green beans, and peach cobbler. After the meal, Ruth cleared the dishes, and the two men sat silently as she brushed the crumbs off the table and brought out her paper and pencil.

With great enthusiasm, she told them about the idea, just as it had come to her, starting with the farming plans, the poultry and dairy businesses, and the leasing of teams.

"And if we all work together, we can make it work. I know we can!"

For a few moments neither man spoke. Ruth looked into their eyes and waited. Finally, Pa leaned back in his chair, rubbed his head and smiled. Ruth hadn't seen him smile like this for months.

"I see sending you to college has paid off after all. You've got a good plan here, Ruthie. Surely couldn't hurt," Pa said. "How do you feel about it, Angus?"

Angus complimented her as well. "I like it. Could work, I'm thinking. It's odd, you know. Just the other day, I was saying to Alfred Dale what a pity it was to see all the old farm equipment just lying waste. I feel it's worth doing, meself. And how will ye ever know, unless ye try?"

Ruth set out the plan to break the project down into three parts: the crops, the dairy herd, and the poultry business. Essential to its success was to start small, and only grow with profits earned.

Angus stood up to return to the Men's Quarters. "After my rounds tomorrow, me and Dale will look over the old equipment, to see what can be done."

Ruth was too excited to fall asleep right away, so she wrote to her college chum, Lois Shaw, now a wife, mother and homemaker in Manchester.

"We're going to put some modern methods of farming into practice, and also new methods of marketing, which I am eagerly investigating. While the methods are new, we're actually going back to the simple system of my grandfather's era, fifty years ago: farming for half the year and logging for the other half. The two activities work hand in hand, the farm providing the food and supplies for the labor force, both men and animals, and the logging providing income when the fields are lying dormant. There are some modern methods I'll research as well.

"It's almost as if I were back in college, studying into the night, but now, this is a struggle for existence! And my ability to execute these ideas has real consequences. Now to sleep, for we rise very early around here, as you must as well, with your little ones wanting to be fed. My love to Winfield and the girls, Ruth."

*Chapter Nine*

**Modern Methods on the Farm**

"Horses for Lease, by Day, Week or Month. Seasoned Logging Teams. Saddle Horses. Reasonable Rates. Contact Ruth A. Park, Warren, New Hampshire," read the advertisement in the newspapers. Within days the first response arrived: a farmer needed a four-horse team to haul logs. Others followed. A lumber wholesaler from Boston, an affable gentleman named Harry B. Stebbins, leased a team to travel to logging camps in the area. A local hotel leased a team to take summer guests on excursions. The leasing business was off to a good start.

"Keep them well-fed and watered. I don't want you bringing them back half- starved." Ruth surprised herself at how frank she could be with the customers. Yet unhealthy horses did no one any good. And she detested the idea of destroying an animal, though she knew owners sometimes had to do that.

True to his word, Angus repaired farm equipment — a seeding machine, a rusted harrow, a blue hay-wagon. From the blacksmith shop in the Lower Farm barn, the sound of his hammer on the anvil rang out, as he wrought out new spokes, handles, blades. Dale, the teamster, readied leather rigging for the horses. Lois Warren Shaw and her husband, Winfield, sent Ruth the money to purchase a piece of machinery for lifting hay from the wagon to the upper haylofts. It looked like a large iron claw that ran along a track in the apex of the barn, to open and release its clutches of hay.

From sales at her farm stand out by the road, Ruth brought in enough to purchase a dozen hens, a rooster, and three pigs. Her next goal was to acquire thirty head of Holstein dairy cows. Within a few seasons the income from the new ventures, selling the milk to the creamery, the egg sales, the vegetable stand, the sale of hay and leasing the teams, began to make a dent in the financial picture. And most importantly, hope and endeavor had

Jason the Bull and Bringing in the Hay.
*Park family collection*

replaced discouragement and passivity.

With Angus's help, she hired men to do the extra work; experienced hands to operate the horse-drawn machinery, to plant crops, make hay, and tend the livestock. Chores included feeding and watering the animals. In the summer kitchen there was a big cast iron kettle where two to three bushels of potatoes were boiled down each day for the pigs. Ruth learned of a new method of mixing cracked corn, oats, and wheat middlings as a nutritious mash for chickens. She purchased 100 pound bags of each, and poured them into large boxes, four feet by four by eight, and then the grains had to be mixed thoroughly with a shovel.

Simpler chores could be done by local youngsters, twelve to fourteen-year-old boys, who were glad to earn money by working summers and weekends. Some would sprout potatoes, (take off the sprouts so the potatoes would not get mushy), endless barrels of them stored in the cool cellar, or clean stalls, and one of the biggest needs was weeding the long rows of cabbages, corn, beans, peas, beets, and carrots.

Operating the farm was physically demanding work, and Ruth measured others' output of energy to her own. She rose early and made a habit of surveying the barns, stables, and crops before anyone had arrived. Then she would delegate maintenance orders, like repairing fences, or wagons, or other equipment. When all was running smoothly, she helped make the deliveries to the creamery.

"One day," she wrote to Richard, "I want to have students from Vassar come experience a summer on the farm. Hard work and fun together. They can sell berries and vegetables at the roadside stand and learn how to do the canning."

The work was demanding, but there were rewards. Swimming, berry-picking, satisfying meals of home-grown food, storytelling, to mention a few. When the crew converged on the favored swimming hole in the Baker River, men and boys and girls and women sported alike, splashing and floating. The rope swing never lacked a rider on these afternoons.

Ruth learned to differentiate between good workers, and those who were not so good. Because the Parks provided a hot meal for the help at noon, and meals cost money, it was not surprising if a young man who started in the morning was let go before lunch.

The new venture was not without trouble. To increase the dairy herd, a bull, named Jason, was brought to the Upper Farm. One day the bull started charging the men keeping him. No one knew what happened but he turned into a furious beast. His bellowing could be heard a mile away. The hired hands tried many methods to subdue Jason, including dousing him with hot water, sticking pitchforks in his side, and even firing a rifle bullet into him, but nothing had any effect. Finally a brave worker deliberately ran in front of the crazed bull to capture his attention. The man ran with all his might for the barn, Jason following close behind. The man scrambled up to a grain feed platform, out of the bull's reach. Meanwhile, two other men slammed the barn doors closed. They were able to chain the bull and end the excitement.

Ruth had to learn about effectively controlling other kinds of rage; her own. An extra farm hand had been hired, and Ruth assigned him the task of mixing the mash in the large bins at the end of the hen houses. In a very short time, the man swung the shovel onto his shoulder and began to walk away. Having done the work many times herself, she knew the grains could not be thoroughly blended so quickly.

"Is this mash mixed all the way through?" Ruth asked him.

"It is," he replied.

She took the shovel and jammed it into the box. Distinct layers of grains could be seen, like gold, brown and tan strata of the earth. "You haven't done this properly," she said, sharply. "This isn't mixed! Look for yourself."

He drew himself up tall and said, "Don't need no blasted woman telling me how to do things." Before she took the time to think about what she was doing, Ruth let a fist fly right into his scowling face.

At that moment, the man's wife, who was helping Mrs. Shortt nearby in the kitchen, came running over to Ruth. "Stop! Leave my husband alone!"

Ruth stormed towards her. "You get back in the kitchen or I'll do the same thing to you!"

The woman, bigger than Ruth in height and weight, retreated, sobbing.

Within seconds, all three people had disappeared from the scene, as if it had never happened: the man slunk into the darkness of the woodshed; the wife, to the kitchen, to be consoled by Mrs. Shortt; Ruth walked out into barnyard, kicking the soil as she walked. Her thoughts were jumbled.

She went to her room, and closed the door. The house was cool compared to the summer heat. Still upset, she turned to the one method she could trust to clarify her thinking, and consequently calm her down: writing in her journal.

*Why did I do that? Why? What makes me act first and think second?*

*What a brat I am. I despise it when I lose control like that.*

She looked at the clock, ticking softly on her bureau. Mother used to say,

"Ruth, count to three before you let temper make you do or say something you will regret."

The incident with the hired hand had occurred in about one and a half minutes, Ruth figured.

*If I had not hit him, and had just held my tongue for a minute and a half, I wouldn't be here now, tormented by my conscience. Couldn't I have waited one and a half minutes before doing or saying anything?*

She thought of Pa, how hard he had worked to make a dream come true; of Joe and Richard, contributing parts of their salaries to educate Esther and Katherine. And her sisters, on their school breaks, kneeling in the sweltering heat in fields, weeding rows of vegetables in the hot weather. Everyone sacrificing for the good of the family.

*That's it!* she said. *I'm giving up my quick temper, and I'll do it by really making an effort to wait out the first minute and a half, before I say or do anything I might regret.*

She named it "The Minute and a Half Rule." Its premise was: Every time I'm tempted to be angry, instead of doing what comes to me to do, I'll refuse to do it, and wait with every bit of my strength for a minute and a half, before deciding what to do. She no longer felt angry at the man for improperly mixing the mash or for his comment about women bosses. Animated with this new view of herself, she found the hired hand in question, and apologized.

"It was the wrong thing to do, and you probably don't believe me, but I've learned my lesson," she said to him. "I'll never do that again."

After a moment, the man accepted her apology, and he said he would be back to his farm duties in the morning. Her mind much relieved, she walked energetically back to the house. On entering the hallway, she looked in the basket where the incoming mail was placed. There was a letter addressed to her from the County Sheriff's Office in Haverhill. As she read the enclosed letter, her eyes opened wide and she let out a cry of delight. It read:

To: Miss Ruth A. Park,

We are happy to inform you that law enforcement officers apprehended Adam Evans six weeks ago in the Brattleboro, Vermont area. He has been convicted of theft, mail fraud, and several other counts. He has been sentenced to twenty years in prison. An association of northern New England timber manufacturers and logging business owners had posted a reward for his arrest and conviction. We therefore are enclosing a check for $200 for the information you provided that led to his arrest. Without the information and evidence you submitted, this would not have been possible.

Sincerely,

Sheriff R.Y. Goodman

She held the letter and check close to her chest for a moment, and closed her eyes. "Thank you," she whispered. Almost more than the financial reward was the satisfaction that she had helped an industry she was beginning to appreciate more each day. She looked at the check again. "I know exactly what I'm going to do with this!"

*Chapter Ten*

**Pacific Vistas**

In March of 1910, Ruth utilized her reward money to travel by train to Los Angeles. Her plan was to meet Joe, who was on leave, and see the sights of the area. Yet an underlying important reason for the journey was to look over property on 7th Avenue, and in Pasadena, land originally purchased by Grandfather Dodge in 1883. (He had invested in the real estate while spending the winter in southern California, under physician's orders.)

Ruth and her sisters and brothers had inherited the California property from Elizabeth Dodge Park, and were undecided as to whether it would sell better "as is" or improved.

Joe had asked a good friend of his, Frank Sullivan, who practiced law in Los Angeles, to look over the property and make recommendations to the Parks. Sullivan corresponded with Ruth, telling of the growing interest in the 7th Avenue district, and made recommendations that the property might be worth developing as commercial property. His research showed two businesses of the future that should be considered: moving picture theatres and automobile repair. The new sights and sounds of southern California thrilled Ruth from the first moment. The balmy air, the white quality of the sunlight as it bounced off the stucco buildings, the shiny motor cars moving along the wide boulevards, lined with tall palm trees. From her hotel room, she looked out on the shimmering, blue Pacific Ocean.

Joe called for her early the next morning in a shiny black Model T Ford, complete with a driver. Ruth enjoyed the ride out to 7th Avenue, where they were to meet Sullivan. As they exited the automobile, Ruth saw a young man approaching with a wave and a smile. He was taller than Joe, with dark straight hair, dark brown eyes, wearing a pin-striped suit, a soft yellow tie, and white collared shirt. He greeted Joe with a hearty handshake, and a warm smile.

Joe, Ruth, and friends, possibly on the 1910 California trip.
*Joe Park collection*

Joe beamed as he introduced Frank to his kid-sister. Ruth had been under the impression Sullivan was a more mature person, not a contemporary. It never occurred to her that Joe was arranging a meeting of this nature. Off-guard, Ruth welcomed the business-like manner in which Frank Sullivan commenced talking of the real estate matters. It gave her a few moments to compose herself. He spoke knowledgeably of the neighborhood, the growth of the district. He showed them around the existing stores, and introduced them to the merchants.

At noon, they dined at a nearby restaurant, and Frank explained the proposals for the theatre and garage. As they waited for their meals, Frank spoke directly to Ruth.

"Joe's told me about the work you are doing in New Hampshire, logging, rejuvenating the farm, and settling the estate. Very admirable, to my mind. How are you holding up through it all?"

"Quite well. Actually, I'm enjoying it. It's good to see the farm not lying waste. And we are making progress. Every penny counts, though, and everything we can do to reduce the debt is paramount. That's why I'm here, to see if it really is wise to look to the future or sell our property here as is."

"Real estate values are going up steadily, as the city expands. You noticed the workers curbing all the sidewalks. Lampposts are going in, and a streetcar line will be extended to this area soon. It's a good piece of property."

He paused.

"By the way, I might say, Joe didn't tell me how good you are with figures. You've kept up with all the profit projections, I've noticed. Your Vassar training?"

"Partially. But college offered very little business training or help in understanding economic issues, on a larger scale. And no practical education on how to negotiate with bankers, and . . . attorneys." She felt her face redden.

Frank laughed. Of course, she noted, Joe would have friends who enjoyed life, just as he does.

As he left to return to his office, Frank told Joe and Ruth he had arranged to take them both to the Balboa Fair the next day. "Breath-taking exhibits, and all the newest inventions — automobiles, telephones, office machinery."

It was true. From the minute they reached the extensive grounds where the fair was being held, the air itself seemed electric with excitement. Women and men fashionably dressed, children running with bags of peanuts and cotton candy.

The afternoon events centered around an air-show. "A Thrill

Every Second — A Monster Flock of the World's Greatest Birdmen" read the posters. At the field, Joe was highly jovial and alert. He watched the show with intensity.

"Curtiss Pusher . . . see it there? Engine created by a motorcycle racer named Glenn Curtiss. Lightweight, fabulous engine. Curtiss is from back east, Ruthie, Hammondsport, New York."

"Where did you learn all this?" she asked.

"The Army is just beginning an Air Corps. I've been learning all I can, studying books, magazines. It's the field I'm meant for, I know. I've requested a transfer. No question, it's the most exciting thing in the military. It's what I've been looking for."

Joe had to leave in the morning to begin his return to Camp Stotsenburg in the Philippines. As Ruth had still another day in Los Angeles, Frank invited her to accompany him on an excursion ride up to the top of nearby Mount Baldy. She accepted.

After Frank had left them at their hotel, Ruth poked Joe in the arm.

"You planned this, didn't you?"

"Why, of course, I did! He's the type of fella' for you. Bright, well-educated, a future ahead of him. No one I'd rather have you meet than Frank. He's top-notch."

*Chapter Eleven*

**Now or Never**

Ruth discovered that it was easy talking to Frank. They had interests in common; literature, politics, business, travel, and music. At the top of Mount Baldy they walked through the meadows and talked about their goals. Frank had plans to move ahead in his law career and make a name for himself. He envisioned a law firm with partners, serving this growing southern California area. Ruth felt comfortable telling him her detailed plan for surviving in the farming and logging life until New Hampshire forestland became revalued.

The breeze moved softly through the tall grass. As they stopped to look at the view, Frank caught her hand, and when she turned to look at him, he gently pulled her close. Ruth didn't say a word, and continued to look up at his eyes. It seemed so natural to be close this way.

"You are a very special woman, Ruth, and not like any I've met before," he said. "I can talk to you so easily, and have intelligent, fun conversations. I enjoy your company. It's odd, but even though I've only met you, I feel through our letters and Joe telling me so much about you, that I've always known you."

Ruth responded, "I feel the same way." It was true, however, whereas Ruth would become quiet when she encountered new feelings, Frank responded in the opposite way.

"I'd like to meet your father, and Esther and Katherine, and Richard. I've heard so much about them all from Joe as well, and now, meeting you, I really want to get to know them."

Before they headed back to the carriage for the ride down the mountain, Frank and Ruth kissed, and held each other in a tender embrace.

"I don't want to let you go away on that train tomorrow," said Frank. "Please promise me that we will correspond regularly and that I can see you next summer when I travel to Boston to my

Joseph Dodge Park
U.S. Army Aviation Corps, 1913.
*Park family collection*

sister's graduation."

On her return to New Hampshire, Ruth reported to the rest of the family all the information regarding the California property, and, with her recommendation, they decided to progress with the construction plans for 7th Avenue, under Frank Sullivan's supervision. Ruth and Frank communicated by letter and telegram, carefully maintaining a professional relationship separate from their personal one.

The following year, in June, when Frank arrived in the east for the graduation, Ruth and Pa met him at the Parker House in Boston for lunch. The two men conversed easily about business in general, and Ruth sat back at one point, beaming with happiness, as she saw Pa getting acquainted with the intelligent, lively young attorney she had come to like so well.

Pa left them alone for a brief time, and they took a streetcar to the Charles River and walked along the banks. With so much catching up to do, and so little time before Frank had to get back to the Sullivan family gathering, their conversation was accelerated and full of laughter. Then Frank grew quiet.

"I have something to ask you, Ruth. Let's sit for a moment." They located a park bench under a tree and he took her hand.

"Ruth, I am in love with you. This whole time we've been apart, I've thought about you constantly. I rush to check the mail for your letters, and re-read them over and over. Look what you've done to my heart. Just to hear your laughter and to be close to you. The thought of being away from you is unbelievable torture. I want to spend my life with you. Ruth Ayer Park, will you marry me?"

The proposal of marriage startled Ruth. She stood up and walked toward the riverbank, so as to not let Frank see her expression. Now the happy flow of conversation of moments ago failed her. Thoughts raced around her mind:

"I enjoy his company, I look forward to being with him, I'm happy to see him getting to know Pa. But marriage? Am I ready to leave all that I know, to live in California? Am I capable of making him a good wife? Can I exchange all that I know for a whole new set of circumstances?"

A clear answer did not emerge. She knew she did not have enough information to make such a decision, but now the question became how to tell him gracefully she needed more time to think about this. For a moment, she was tempted to condemn herself for not being able to trust her heart in such matters; not leaping at a opportunity most women were longing for, the opportunity to be a wife, a homemaker, and a mother. She thought of her geometry professor. "Collect and record all possible information. Then, proceed to solve the problem." She went back to the bench, and sat down. "I need time to think about this, Frank. I have a great deal on my mind right now with the farm, the lumber business and the estate to settle. You are the bright spot in my life, and I cherish our visits and letters more than you could imagine. I want to make a fair decision for all concerned. May I ask for some time to think this through?"

"I know what you are concerned about. Your responsibilities to the family, to your father, to the estate. I can arrange to come back in the fall and help you close down the farm, sell the livestock, and pay off the help. We can put the property in the hands of a reputable real estate manager. Your father will be with us and we'll start a new life, together, in Los Angeles. I will be a faithful, good husband to you."

How she loved his persuasive manner, his sincerity, and his good intentions. This was truly an exceptional man. Although her heart could not be convinced that she would be happy in California, here was an individual she knew would never do anything to hurt her.

"Frank, I need at least a year to see if I can sell some of these New Hampshire properties and settle my mother's estate. Then I will know if I will feel free to marry and move away." Reluctantly, but lovingly, Frank agreed to wait a year for her answer. They sat quietly holding hands and watched the boats on the river, until it was time for him to leave. He escorted her back to the Parker House, where Pa was waiting in the lobby.

In July, Joe was accepted into the U.S. Army's Signal Corps and received training at Hammondsport, New York, and then was transferred to San Diego, California. He and Frank visited on occasion, and often the conversation turned to Ruth and life in New Hampshire. Tempted as he was to interfere with his younger sister's decision to put marriage on hold, Joe never said a word to her about it, either in his visits home or in his letters.

It was in San Diego that Joe finally qualified for his aviator's rating on February 15, 1913. He wrote the family of his love of flying, and his early success at setting aviation records; the first to fly at 8000 feet, the first to fly 50 miles over water and the first to fly at night. He wrote that he was eagerly training for a pioneer solo flight between San Diego and Los Angeles.

One early morning in April fire broke out in the barn at the Park Upper Farm, caused by a fallen lantern. Ruth rang the big fire bell that hung on a pole in the yard. Angus, Albert, and other hands from the Lower Farm were first to arrive to help Ruth and Pa.

"Get the animals out of the barn!" Ruth shouted.

Angus rushed to open the barn doors and plunged into the smoky darkness to open the stalls. Several other men directed the fleeing horses to the open fields, swatting the flanks of the animals as they passed.

"Open the pig-pen!" In the confusion, some of the freed animals turned and rushed back into the fire. Two horses and two pigs perished.

Ruth quickly settled Pa in a big armchair that had been brought out by one of the farm workers. Then she and others made repeated trips into the house to remove whatever they could.

A horse-drawn fire wagon arrived, but the muddy fields delayed it in getting to the fire. The house soon caught fire as well. Nearly a hundred people were on the scene, and for one hour all fought bravely, but to no avail. The house, barn and out buildings were lost.

Ruth wrote to her Vassar friends: "Then came a new experience; we moved down into an old house where the help had been kept. . . . We lived in two rooms by ourselves. . . . Outside were the old hay barns, the cattle tied in makeshift stanchions, open drains and the pigs reigning supreme in a hurriedly made pen in the front yard! Then came pandemonium, the logs to be got in, the cows to be milked, the barn to house hay and stock to be built and the old ones to be torn down as fast as the new one went up."

Two weeks later, on May 9th, Ruth's twenty-eighth birthday, news came from California that Joe had lost his life during the flight from San Diego to Los Angeles.

During the next few days, the story of the mishap unfolded: Joe's mechanic had gone by train to Los Angeles, where he and others including Frank, would be on hand to congratulate Joe upon his arrival. Along the coast, Joe encountered fog, typical of that region, and landed near the town of Olive to get his bearings. After talking with people there, and asking them to telephone the delay to his superiors, he re-entered the craft and began the take-off down an alfalfa field. The craft failed to gain enough speed to get airborne and went down into a ravine at the end of the field. Joe didn't survive the crash.

Frank accompanied the body back to New Hampshire. After the service, the casket, draped with an American flag, was transported to the Holderness Trinity Cemetery, where the Dodge plot is situated. As the procession passed the Holderness School, the preparatory school Joe had attended before West Point, a bell was tolled in his honor.

Frank stayed on for several days, helping with arrangements and lending a steadying influence to the family. He asked Ruth to join him for dinner at the Pemigewasset Hotel in Plymouth, adjacent to the train station. In several days, there had been not much time to speak privately.

"Ruth, when we planned to meet again, I never dreamed that these would be the circumstances. But I must admit to you, that when Joe told me about this mission, I could not help but see the risks involved. With his permission, I placed the properties he owned in Washington state in a trust for his three siblings. Of course, I can help you in the legal aspects of his estate."

The candle on their table flickered, and shadows danced on Frank's face. His eyes earnestly searched Ruth's. She leaned forward, and tenderly took his hand.

"Thank you so much, Frank, for all you do for our family, for your help in all matters. I owe you an answer, and although it seems not to be a suitable time to discuss this, it's probably the best, because I can see clearly now that it's not possible for me to marry you, Frank. My place is here, with Pa and the farm and the company. These hills are my home, my roots, my heritage. I belong here. We are in different worlds, and we are needed, each in our own world. Can't you see? To leave would be impossible."

Frank sat back in his chair, and his hands fell to his lap.

"Oh, Ruth, Ruth, Ruth, isn't there something I can do to change your mind?"

"I'm sorry, Frank. My answer is no."

For a moment, she wanted to soften her answer, in some way making it easier for him. But she respected him too much to do that. Experience had shown her the wisdom in leaving a person alone to solve his own hurt, and trust he's strong enough to do it without sympathy. Although they would retain the business aspect of the relationship, as attorney and client, when they parted, Ruth realized the relationship was changed forever. It was the most difficult good-bye Ruth had ever had to say.

As she rode the train back home to Warren, she reflected on her decision. "I must be out of my mind!" she thought. She thought of how happy Joe had been to introduce Frank to Ruth, and to see the friendship grow. And yet, in her heart, she knew that her place was not in Los Angeles, and to ask Frank to exchange all that he had worked for to live in the woods of central New Hampshire would be cruel.

One thing, she was learning. Decisions get tougher. Only a few years ago, her parents and brothers influenced most of her choices. Now, what she made of her life was up to her.

Twitching the logs out.
*Park family collection*

*Chapter Twelve*

## High Gear

Joe's death wakened Ruth to the urgency of the tasks before her. Grief would, outwardly at least, have no room. The time to work was now. She could think of no greater tribute to Joe than to destroy once and for all the haunting debt (the ghost of past failures) and set the course right for future Park endeavors.

Simeon Lyeman, the bank president in Laconia, wrote: "The situation with the Park indebtedness is absolutely pressing. We are exceedingly reluctant to carry the indebtedness any further with only partial and sporadic payments."

Ruth cried in disbelief when she received this. "What a lie! Except for these last three months, payments were made in a timely fashion."

After the fire, Ruth desperately needed to shelter the animals, and to improve the quarters for Pa and herself. The insurance settlement would take several weeks and so Ruth saw no other choice than to use the mortgage payment to begin rebuilding the barns. When she notified Lyeman of the situation, he had declared he would "look the other way" if her payments for two months were delinquent. But when the insurance check did arrive, so did the property tax bills, and Ruth paid back the two late payments to the bank, and used the rest of the monies to pay the taxes. This put her behind again on the monthly payment to the bank.

Lyeman's next letter warned, "No more advances on logging operations, as they have invariably proven unsuccessful. Money is very scarce, and we will tolerate no further considerations in view of past results."

Lyeman had copies of these letters sent to Richard, urging him "to take charge of things, and assure some of the creditors that somebody, in authority, can go ahead and turn some of the property, somewhere, into cash. With your father sick, and situation growing worse, it might result in assignment, or foreclosure."

Unhappily, the economy in southern California took a sudden downturn, and the Los Angeles properties stood vacant. Lyeman's third letter made no attempts at being civil.

"Bring the account up to date, or we begin foreclosure proceedings."

Richard came to Ruth's defense. He wrote Lyeman,

"Referring to your letters of late I am sorry that you feel that my father's and sister's finances are in such bad shape. . . . My sister has the sole handling of the N.H. property and her local judgement in her affairs is far better than my long distance advice. I know that she and my father are in fact trying hard to liquidate their real estate, and I hope that you as an old friend of the family will not take any steps looking toward foreclosure until it becomes de facto necessary to protect your bank, and you must admit that that is a long way off. I regret that my calling makes it impossible for me to give my home folks any personal help. . . . am going to Los Angeles in the hope of straightening out matters there. Very sincerely, Richard Park "

Ruth refused to show any of this correspondence to Pa, or to talk to him about the bank's threats, for the discussion would most certainly put him into a state of despair.

But unbeknownst to Ruth, in the summer of 1917, Pa secured a loan of $30,000 from the timber wholesaler, Harry Stebbins, of Boston.

"I'm buying a tract over in West Rumney," Pa announced one night at supper.

Ruth couldn't believe her ears. "What?"

"I'm going to log it, and by spring, I'll have enough to pay Stebbins back, and pay off the mortgages on these New Hampshire properties," Pa replied.

"Pa, how could you? You're defeating the whole purpose of what we're working for. That's way too big an obligation. We all decided to make it back the new way — small logging jobs, leasing the teams, hay, and farming profits. Another large obligation, like this, will put us under, Pa. Please, let's talk about this!"

"It's too late. I've made up my mind. I've got the deposit on the land, and I've ordered the equipment and supplies. All we need now is to find a portable mill to lease for the season, and hire on additional crew. You'll see, we'll make it a 100,000 board feet per week operation."

"Pa, you can't do this! It's not fair! It's just what we shouldn't do. I beg you, Pa, don't do this!"

"Young lady, just remember who you are talking to. I'm the

father around here, last I remember, and I don't take orders from you." His shoulders and hands trembled.

"Pa, you don't have to prove anything to anyone. You didn't fail. The lumber business reached its peak back in '04 and you got caught, over-invested, at the wrong time. It could have happened to anyone. You don't have to make it up to anyone. Let's give the money back. It's not too late."

Pa's face reddened. Storm clouds gathered in his brow.

"I don't want to hear anymore about this. I'm still the head of this family. You're not the only one with ideas."

For the next three days, whenever Ruth would speak to him, he would turn away in silence.

Ruth's heart was breaking. She knew the scheme wouldn't work. After twelve years' experience dealing with paper companies, local mills, wholesalers, and other loggers, she knew the time was not right for a big venture.

"Pa's ideas are always outstripping his execution," she wrote to Richard. "I've no choice but to go along, although I know it can't succeed."

Late in the season, Ruth secured one of the last portable mills in the area. The war in Europe was underway, and Finns Agency in Boston found few men strong enough or skilled enough to send to the woods.

As each day passed, and spring approached, Ruth's calculations showed the operation would produce barely enough to pay the help, the lease for the mill, and only the loan payment for the equipment. The promise to Stebbins for full repayment by spring could not be kept.

Downcast, Pa returned to the farm, and Ruth arranged for his care, and checked on him frequently. This was possible, as she followed the wagon loads of logs to the mill. (She was determined to verify the scaler's measurement of each load.) On her visits to see Pa, she would bring encouraging accounts of the day's events.

"Beautiful trees found in the back part of the West Rumney lot. Tall pines with easily three sixteen foot logs in each of them." Or "I learned today Parker Young Company's paying good money this month for pulp."

Nothing seemed to make a difference. He just sat, hunched over, eyes staring straight ahead. The doctors could do nothing to cure the despondency.

The winter brought other sad news as well. Ruth's younger sister, Esther Park Ward, passed on. Though a doctor herself, and married to a physician named John Ward, no efforts could save her from complications related to childbirth. She died nine days after a daughter, Esther Ward, was born.

With Richard and Katherine so far away, (Katherine had married and moved to the Philadelphia area) Ruth could herself have settled into despondency. But she reasoned, "What good would I be to my family or employees, if I became ill or without hope? Self-pity is a luxury I cannot afford." So when Simeon Lyeman's letters began again, whereas in the past his tone had undermined her self-confidence, now it charged her with determination. She relied completely on the cure that had always held her in good stead: hard work towards an honest endeavor. Whatever the season, or the weather, there was always work to be done. She wrote to the Vassar Alumnae of the "pleasure in the woods work that the human element could not dim. I have always loved it; the good smell of the pines, the sound of axes and the crash of falling trees, the creak of the big teams swinging down the mountain side, and through it all an air like wine and a sparkling sunlight. It is intoxicating. One feels the perfect rhythm in one's body."

Whatever circumstances developed, she felt she was ready.

*Chapter Thirteen*

**Retreat**

There was nothing that could be done about the West Rumney venture except to go back to Ruth's methods — attacking small logging jobs, with limited help, and being vigilant each day to the changing lumber economy. For now, that meant drawing out of the woods loads of pulpwood, four-foot logs suitable for the paper mill's vats.

Stebbins was a stable, reasonable business man, and he agreed to take monthly payments, at least for a period of time. Ruth oversaw the operations from the West Rumney camp, high up on the ridge overlooking the Baker valley.

Back in Warren, Mrs. Shortt cooked and kept house for Pa. Three hired hands from town took care of the animals while Angus, Alfred and Ruth were away in the woods. In between jobs, Ruth returned to the farm, where her days were spent visiting with Pa and solving problems regarding the help or the livestock. In the evening, she read. As each year had passed, her appetite for good literature had increased. She feasted on Joseph Conrad's stories of the sea, the thrilling descriptions of flying by pilot Antoine de Saint Exupery, along with other authors, poetry and current events magazines.

In a letter to Richard, she wrote of de Saint Exupery as one who could write "like an alchemist, with the power of fusing ideas into the original pure element. And his medium is the written word."

Ruth wanted to write about life in the lumber woods. One winter evening, when all the help was off the West Rumney project and temporarily back at the farm, she retired to her small room to read. Through the thin partition, she could hear Angus and Alfred conversing about the early days of logging, when Angus had handled large crews in the White Mountains, and Alfred was a teamster, hauling towering sled loads of logs in the

Will Park and family members
ready for a swim, circa 1920.
*Park family collection*

Lincoln area.

The two woodsmen were resting comfortably around the potbelly stove in the Men's Quarters. As Ruth listened to the rich tones of Scottish and Irish accents, she realized here, right in front of her, was a valuable opportunity to record the story of the lumber woods, and at the same time, develop her writing skills. She moved her chair quietly nearer to the wall, so she could hear more distinctly, and began to copy down the conversation of the two woodsmen. In all, she wrote down five amusing stories, capturing as much of the dialect and colorful language as possible. When she went to visit Katherine in Germantown one winter, she borrowed a typewriter and typed them up, and entitled them "Lumberjack Tales." She sent a copy off to Richard, who responded immediately, "Good work, Ruthie. Vivid descriptions of a life of long ago — when men were men, as Angus would say."

The diversion of this literary effort perfectly counterbalanced her out door work. She began to enjoy a deeper dimension of life, as this creativity enriched tender shoots of an inner peace that she had not known before.

In October of 1920, Pa was up at the old homestead on Cummins Pond, working on the engine of his automobile, when he suddenly collapsed. Ruth got medical help for him, and stayed close. He passed on a few days later. He was sixty-four.

For fourteen years, they had provided for each other the sense of family. Both strong-willed and opinionated at times, they wrangled out the daily problems of life with humor, forgiveness and consideration of each other's feelings.

A few months later, in January of '21, while Ruth was away at the lumber camp, and only a few hands were at the farm, a second fire erupted in Park Flat. This time the cause was a faulty flue at the Lower Farm house. The 45 mile per hour winds and 5 degree below zero temperature worked against the efforts of the firefighters.

The *Warren News* reported, "The big engine was rushed to the scene behind four heavy horses, the State truck with men and equipment, trailing the hose wagon. The engine was set on the bank of Ore Hill Brook, across from the farm buildings. For 5 hours, the water was pumped across the meadow drowning out the fire but filling the cellar full of water. The fire was held to the ell and did not gain six feet from where it originated, but the main part of the house was destroyed as firemen held the flames from the large barn with stock and thoroughbred Holstein herd. . . . Connected to the barn is one of the largest silos in the state. . . .

Two cars and farm equipment were saved from the shed between the house and the barn. None of the furniture or personal belongings could be saved."

Ruth later wrote of the experience this way. "As I reached the ridge at the head of my lovely valley I could see nothing but flame. It was one of those steel-blue January nights of a stinging cold far below zero, but as I reached that ridge a mile from the farm, the air was hot against my cheek. . . . Our efforts could amount to nothing."

The losses included family mementos. Boxes of Mother's art supplies, several of her drawings, her wedding dress, the family china. The grief of losing Pa, in combination with the new challenges, threatened to overwhelm Ruth. Her heart and hope were broken with a desolation that she had never known before. How she wished she had given those family heirlooms to Katherine or Richard. Though Ruth's life was barren of many material comforts, the few treasures she did have had meant more to her than she realized, and she felt unspeakable sadness that they had been destroyed, faulting herself for being away from home so often.

Unfamiliar with this new experience, and frightened by her inability to make sense of it, she instinctively knew she must seek solitude and regain strength. She thought of the serenity of the West Rumney camp. It was only ten miles away; the buildings were rough, but adequate.

The next day, she made her decision. Bringing only Angus into her confidence, she said,

"Angus, we must provide shelter for those who will stay. Buy canvas and we'll make tents with wooden floors close by the ell. As soon as these are done, I'm moving up to the little shack at the edge of the woods, up at the camp. Rebuilding the farm will have to wait."

For two years, she concentrated only on the lumber work, focusing all her attention on keeping Lyeman from foreclosing. Her tasks on the lumber jobs were those of any business owner: do whatever needs to be done, and especially the things no one else wants to do. She scaled the logs in the yarding area. She followed each load to the train yard or mill. She checked on the cutting work, and the logging horses. If rigging or equipment needed repair she would often bring the items to Angus to fix. Toting in food and supplies to the camp gave her the time to stop in the local libraries to replenish her reading supply.

The hard work stilled the sadness within. But in the evenings, her heart ached. Like a scientist in a laboratory, she examined the phenomenon of her plight and experimented with solutions. Two things soothed her, as never before, although they really were one thing: beauty; beauty reflected in art, especially, in literature and in nature.

In the quiet twilight, with her Newfoundland pup, Sancho II, guarding his wounded princess, she contemplated the beauty of the wilderness around her. And as darkness fell, she read or wrote. Her love of writing, too, comforted her. Faithfully, she communicated with Katherine and Richard, outwardly on estate matters, but inwardly, as two of the closest friends, whom she had grown up with and loved.

"I am down at West Rumney at the camp. Really a lovely place. From the little house I sleep in I can look out onto the village in the valley, over treetops, over onto three tiers of the bluest of hills. . . . so quickly do they rise they really seem like mountains. Their line is lovely and the most enchanting music of veerys and thrushes from the woods in back of us. Every evening and morning and late afternoon gives me a musical festival every day that symphony lovers would have to turn green with envy if they were ever mice up here. So quickly surrounded am I by woods and swiftly rising hills full of music, that I really feel quite content — until the whip-poor-will comes and goes in back telling all the little birds to stop their singing and go to sleep. Then not a sound (except) occasionally when some youngster thrush, braver than the others, breaks the utter stillness with one perfect echo of his song. Really quite nice after all, made nicer by a swim in the river just before sunset.

"Then after the thrushes are quiet and even the bustling, bustling night policeman of a whip-poor-will has stopped his assiduous efforts after "poor will," you look out your window and see all the fireflies' lights go open and shut, now here, and now way over there, you'd think all the little stars had fallen down and gone a dancing on the ground. Then to a nice long sleep not a sound but the whir of the sawmill nearby, hurrying to finish their cut. "Poor Will" is just catching it . . . . no I think it is Will Thrush — one perfect little song. Yes, it is quite nice and I am really not discontented at all at such times. . . . The smoke from that mill looks just like a lake over in the hills. It is too dark to write anymore, even with the clear moon. . . . It is bewitching, these soft lovely hills, melting into the sunset colors, still lingering. Pretty soon, it will be midnight."

To her college friends, she continues the account:
"With this song still in my ears, I went back to the farm,

resolved to find a way to farm and still keep well rounded my love for beauty, whatever form it might take in a farmer's life. And here I have been, rebuilding while I farmed as I could. The old farmers regard me as a riddle, but the children accept me on my own ground. I'm still tackling the problem, but seem to see a glimmer ahead. Tomorrow, I shall see the first delicate haze of the green and red of the maple and poplar buds beginning to spread like a veil over the pasture side. Yes, I've something worth it all."

## Chapter Fourteen

### Home Free

To celebrate her rejuvenation of spirit and her return to the farm, Ruth went to a beauty salon and had her long hair cut into a fashionable bob. She wrote on the bottom of a letter to Richard and Katherine, "I've bobbed my hair! I've bobbed my hair! I've bobbed my hair!" In contradistinction to Samson who lost his strength by a haircut, Ruth gained energy from the new look.

As in the past, ideas and plans for the future filled her mind. An encounter with lawmakers at the Capital in Concord regarding logging breathed new life into her desire to be a legislator. Travel to Mexico tempted her. She corresponded with Katherine on plans to help with Maskoma Lodge, the summer camp that Katherine had established on the family homestead in Dorchester.

But Ruth saw clearly that these activities would have to wait. The farmhouse and barn had to be rebuilt. More importantly, the cost of being in debt was compounding, and it infuriated Ruth to pay penalties and extra charges. Never had she felt more determined to be debt-free.

One day, while returning books to the Plymouth library, she found a quiet corner with a large table and comfortable armchair, and sat down. Her whole being ached for a clear answer to the financial troubles she'd faced for so long.

She placed a notebook of lined paper in front of her and sat still, hands in her lap.

"There MUST be a way! There MUST be. How can all this good work of three generations be for naught? We have built a business, and learned it well, and we have paid dearly for this knowledge by our errors, and now, can there be nothing but endless trading of dollars with no progress, and the real threat of losing it all through default and tax sales?"

"Please take the blindfolds off my eyes," Ruth prayed.

An idea came to write down a list of all the assets of the estate.

Jobildunc Ravine, Mt. Moosilauke, 1896.
*Ellen Webster*

The buildings, the farmlands, the stock, the equipment and the woodlands. All the property had at one time been listed for sale with real estate agents, and none had sold. In the past, she had requested that the agents show the properties to prospective buyers, feeling herself to be unskilled in sales techniques. Agents she had hired in the past had not known the terrain and had taken prospects out into the woods and most of them had gotten lost in the thicket, never seeing the stands of good timber. The thought came, "No one knows these properties like I do. What I lack in salesmanship, I have in abundance when it comes to love of the land, and understanding of the objections a buyer might wrestle with, and, most of all, I can visualize the possibilities."

But another inner voice said, "Me, persuade a buyer? I couldn't do it."

Yet, she was desperate. Her eye scanned the list again, and stopped at the 3000 acre tract on Mount Moosilauke in Jobildunc Ravine, the rugged area named after three men — Joe, Bill, and Duncan. The pulp market was improving now, and recently land in that area had sold to paper companies.

"I DO know that land well-enough. I could lead them."

As a girl, when W.R. Park was logging the area with the gravity railroad and three lumber camps on the slopes, she had spent numerous weeks out in the woods with her father and brothers laying out access roads.

More recently, she and Katherine had led expeditions of young campers from Maskoma Lodge on hikes and overnights in those same woods. Suddenly, she was on her feet.

"I could do it!" she said, right out loud.

She closed the notebook, and walked outside to her truck, and drove home, deep in thought. She knew all that the task at hand would entail to show the land at its best to a prospective buyer: waiting for deep snow to cover rocks and smooth out the rugged terrain, rising at four to have a long-enough day to show all the property, and snowshoeing along carefully chosen routes past the best stands of the virgin red spruce.

In preparation for the work ahead, Ruth spent long hours in Jobildunc Ravine planning how she would guide buyers to the timber, and preparing a map.

In early fall Ruth advertised the property. Word reached interested parties, telegrams and letters arrived, and Ruth made appointments to show the land. A longtime friend in the real estate business steered some referrals her way, as well.

The government agents were first to come and look. After the arduous tramps on Mount Moosilauke, the negotiations began, and soon confirmed what Ruth had already known. They would pay so little! It could not meet the Parks' need.

Ruth was at first disappointed, but one day in November of 1922 word came that the Parker-Young Company of Lincoln wanted to see the tract. Ruth's senses were keen as she planned this appointment. She could make no mistakes. This was the "meanest, shrewdest company in the east," she wrote to Richard, whose agents needed large quantities of pulpwood to feed the hungry digesters at their mills.

All her efforts had brought her to this point, and now was her opportunity to sell the land to a company which could complete the logging, with a parcel that would round out its existing holdings on the mountain, and who would see the possibilities for future growth and harvesting.

The Parker-Young agent was a man named Richardson, an experienced woodlands manager, older than Ruth by about twenty years. He brought with him a new employee that the company was grooming for woods manager, a Dartmouth graduate.

At dawn, the three started out on snowshoes, easing carefully across the headwaters of the Baker River, only partially covered with ice, showing gaping holes and black swirling water. Then they began the ascent along Gorge Brook, Ruth leading the way.

By mid-morning, they had bushwhacked up to tree line on East Peak. It was a cloudless day, the temperature near zero and the wind roaring. Old Baldy, as the Abenaki tribes called 4800 foot Mount Moosilauke, towered above the surrounding terrain and was unprotected from the wind. To talk or look at a map, the three had to crouch in a clump of firs.

"Can't stay up here long," hollered Ruth. "That wind feels like a few minutes ago, it was chilling itself along the Great Lakes."

The young man nodded in agreement. Ruth saw his brow wrinkled with concern. He said to Ruth,

"But are you sure you don't want to rest a bit?" Ruth laughed and shook her head. Despite her short stature, she hadn't met a woodsman yet who could out-snowshoe her. Richardson and his assistant were the third group that she had led up the mountain this month.

After a cold lunch behind the windbreak, the trio crossed East Peak for a look down into Jobildunc Ravine. Ruth pointed out the boundaries, first on the land, and then by referring them to the map.

"And there," she pointed towards the upper part of the ravine, "is the best stand of spruce you'll find anywhere. My father tried his darnedest to get those down, over fifteen years ago."

"With today's equipment, it could be done," offered the assistant.

Richardson didn't say anything. He instead surveyed each side of the Ravine. It was obvious that years of logging old growth spruce had made him reverent. He said nothing.

Ruth led the men on a careful descent into Jobildunc Ravine and then for the rest of the day marched them by as many huge spruce as she could. As Ruth had pre-arranged, in the late afternoon the threesome made their way to a cabin in Will Park's old Camp #3, now owned and operated by Champlain Realty, a subsidiary for International Paper Company. The hut was well-equipped with food, firewood and supplies.

After an early dinner, the two men retired to the bunkroom, and Ruth took the couch in the main room, near the big wood stove. She stayed up awhile to read, and write to Katherine.

"Dear K, Here I am away up at the foot of Jobildunc in a log cabin of white birch logs, face to the big spruces on the other side of the brook: the air filled with the laziest of snowflakes. Jobildunc is still frozen up solid. You could hear him grunt when you tweaked the little spruce too hard on the top of his chin. Little does he realize that maybe another year they'll be pulling his whiskers and pulling his hair by the handfuls until he IS bald on the top of his head as an eagle, and driving them down the river pell mell to the big city.

"Cutest little office you ever did see — I have a great fire in the box stove, Conrad's sea tale at hand and the New Republic, . . . staying tonight in the Realty camp with Mr. Richardson and a man — both sent up by Parker-Young Co. to look over timber. I showed up Jobildunc in first class fashion — kept out of all the thickets and in timber all the time."

In December, after weeks of waiting, Ruth received not only the report, but a purchase and sales offer from Parker-Young. She sat at the kitchen table in the farmhouse, reading the contract over and over. It was too good to be true. The amount would not only pay off the back taxes and mortgages at the Laconia bank, but there'd be some money left over for Katherine, Richard, and Ruth herself.

It was a moment to savor. Ruth reflected on the years of struggle, the fighting for each dime, the sacrifices, the heartaches. She wished her father could have been there to hear the good news. Soon, she would go to the telegraph office at the railroad station to notify Richard and Katherine of the offer. But for another few moments she simply sat, enjoying the fulfillment and peace. She thought she'd never known such happiness.

"Thank you, thank you, thank you!" she said, in a prayer of gratitude.

The negotiations went quickly and, in a matter of weeks, Ruth had Parker-Young's check in her hand. She drove the Diamond Reo truck to the bank in Laconia, and at a teller's window, she arranged for the money to be transferred to Richard and Katherine's banks. However, the portion for Simeon Lyeman, she asked for in cash.

With the money in a paper bag, she waited patiently to be let into Lyeman's office. She smoothed her hair with her hand, and brushed a bit of sawdust off her wool trousers. Her boots looked scuffed, but she didn't mind. When announced, she immediately rose, and walked passed the secretary, and placed the bag on his desk. The banker's eyes grew round with surprise.

"Here's the money we owe you. All of it," she said, calmly.

His mouth gaped, as he fumbled for words. "Why, why, you don't say?"

"Yes, I do say. It's all here," she responded, as she confidently moved the ornaments on the polished mahogany desk out of the way, and began making piles of bills.

"Some of your real estate sold, did it?" he surmised.

"That is right," she said. "May I please have a receipt?"

"Why, certainly, certainly." He motioned his secretary to assist him.

"It's funny," Ruth thought, as she sat, watching the two of them count the money. "He doesn't look like such a giant anymore. Just kind of ordinary and a little pitiful in his over-grand clothes and stuffy office. Wonder how I was ever afraid of him?"

As she drove home to Warren, she thought, "I don't have a wealthy husband, or a fancy home, or a big bank account. But what I've got right now, I wouldn't trade for anything. Peace of mind, my pride, and a feeling of hope for the future. There's no price tag on these. Without a doubt, this is a high point of my life right now, and I'm one of the happiest people on earth."

The rest of the drive, she sang all the songs she could remember. She waved and smiled at people along the way. "That's right! I've got something good after all."

The river, the valley, the mountains, the sky had never looked so beautiful.

"I'm glad to be alive," she said, out loud, "and I'm never going to forget this feeling, ever."

She honked the horn at a passing lumber truck, and patted Sancho II, and exclaimed, "Let's go, Diamond Reo. Get me home in time for the sunset."

## THE END

# PART III

Lumberjack Tales

Written by Ruth Ayer Park

Two Baker River valley woodsmen,
Alfred Dale and Freenan King.
*Photo courtesy of Zola Ostrander*

# Lumberjack Tales

*Author's Note: Ruth's roughly typed manuscript, probably from the 1930s, was retyped by her nephew Joe Park in the mid-1990s. This was done in order to smooth it out, and make it more readable for transcribing. The tales are essentially unedited. Here you are presented with just what Ruth wrote down, as she listened from her side of the thin partition that separated her father's and her rooms from the Men's Quarters. The rough quality of the two lumberjacks' speech has been only slightly altered.*

Angus McMaster and Alfred Dale have worked for me in the woods on the small jobs I have carried thru at various times, Angus steadily for the past twelve years. Angus is a typical Scotchman, dour as the worst old Scot that ever lived at times. Let the wind blow pleasantly, however, and his humor and naturally merry but profane temperament are thoroughly enjoyable. He has spent his life in the woods ever since he left Nova Scotia, taking charge of large camps in his prime. He must be nearing sixty-three now though he declares he is only fifty-eight. But still a muscular hard-working capable man, swift with his axe as well as his tongue, quick tempered and generous.

Alfred Dale often comes to work for me, but never stays long at a time. He is a roly-poly Irish wit with a penchant for married ladies provided they can make good doughnuts and pie. Alfred doesn't care much about how much meat his different boarding house mistresses supply but he insists on good pastry. While his affections are always purely platonic, he prefers a complacent husband not too addicted to home brew, for it has been known to happen that his relations were misinterpreted. In fact, one or two husbands have been known to throw Alfred out-of-doors. However, Alfred has never been at a loss for a boarding-place when woods work was slack. Many a door is open to him still especially after the drive when his pockets are full.

But last winter Alfred hired out to me to do chores. All kinds of work was slack everywhere. I suspect that his board was considerably in arrears and the last husband was inclined to look askance on a continually empty pocket. So I hired Mr. Dale until times should look up a bit in the woods. The only other help was Angus and these old lumberjacks would sit toasting over the men's room stove swapping tales and lamenting the old days when men were men, in the woods or on their seasonal sprees.

## P.I. Kelly and the Monk

Many a time I would hear Angus saying, "It ain't like the old days." And one evening I remember particularly. He was saying, "Why, Alfred, these little shrimps they send up now into the woods ain't wuth their salt. They ain't got no muscle and no brains neither, the bunch o' truck-drivers and ditch diggers and waiters and then they expect a man to get logs out o' the woods."

"That's right, Angus," Mr. Dale's mournful voice would agree. "It sure ain't like the days when Luke Reagan and Andy McAllister and you and me was up in Wind River, and P.I. Kelly. Why, that P.I. Kelly, I never seed a better man with a cant-dog. Saw him in Boston just a month ago, and say there he was out beggin' drinks and that doubled up with rheumatism, talkin' 'bout going back to 'Skish'."

"You don't say," replied Angus. "Why don't seem like it was longer' n one winter ago, P.I. Kelly and me was down to Boston, atryin' to see how much we cu'd git for our money. P.I. Kelly, he sure was a great lad. Had a face like the inside o' a watermelon and y' cud run circles through his legs. Used t' say he got them bow-legs ridin' broncos in th' west. 'P.I.' I used t' say when he'd begin some o' them yarns o' his, 'you never got no nearer West than the New York line, and I'd lay a dollar y' never rode nothin' but a buckin' log.' P.I.'d get pretty mad at that and say anyway it took more nerve to ride a buckin' log than any bronc he'd ever rid on. He had his nerve, P.I. did. I'll never forget that time we took that vacation together. We'd been up at J.E. Henry's, richest old rascal ever in the lumber woods and smartest to make money. What ef he did take it into his head to make money off the insurance companies now and then."

"But as I was sayin', P.I. got a stake a dollar a day he was workin' for then, and sez P.I. to the old man, 'I guess I'll be goin' out,' and sez I, steppin' out from behind P.I. 'Guess I'll be goin' too, Mister Henry, for a few days.' The clerk he rolls out a good three months check and we beat it to Boston on the freight to save that check till we got to where the good likker flowed."

"Well, as I was sayin', we got that likker alright. Then one fine mornin' P.I. woke up and felt in his pants pocket for the rest of the roll. But there want nothin' there but a piece o' chewin' plug and one nickel. Well I done the same and b'god nothin' but a plug o' B.L."

"'Well,' says P.I., 'if that don't beat it all. Angus do you remember what we done last night?'

"But fer the life o' me I can't remember a ---- thing."

"Well," says P.I.," Angus, we sure can't finish this vacation. Guess we'll have to hit it back to the Old Man's, 150 miles and only a nickel between us.'

"I just grunted, my head was that big. I felt as if I'd been plugged anyway. P.I. he opens the winder and looks out for a breath o' air. In through the winder comes the sound of an organ grinder. I gets up and leans over P.I., and there's a man down on the street playin' to quite a bit o' people, and a monk a passin' the hat. As fast as one of the folks listenin' to the music would flash a nickel at th' monk, he'd jabber a bit, then dart over and grab the nickel and into the hat it'd go, 'fore y' could say Jack Robinson. I never seen a smarter monk."

"Well, P.I. nudges me. 'By ! ! !' says he, 'Angus I got an idee,' and he pops a nickel, our last nickel by ! ! !, out o' his pocket and leans out o' th' winder and flashes that nickel at the monk down below, three stories in the street, and jabbers to 'tract that monk's attention. The monk, he sees the nickel and the man, he says, 'All rightee, gettee tha nick, monk.' Up comes the monk leg over leg up the wall and sets himself on th' winder-sill a holdin' out his bag."

"Well, now, I don't hold to have had nothin' to do with such downright piece o' thievery, but if that P.I. didn't snatch that bag right out o' that monk's paws without even giving th' nickel and then he makes a dash for th' door with that monk a jabbering on the winder-sill and the man down below a yellin', 'P'leece, pleece, come quick, steal — a my mon.'

"Well as I said, I don't hold with no such downright thievery as what that P.I. done to that pore little monk, but I just concluded that want no place for me, and I lit out after P.I. down the back way and up an alley, and say, the last I seen o' that P.I. was a turnin' a corner, and I, I just sat down where I wuz, my head was that dizzy a crackin'. No sirree, believe me or not, I never set eyes on P.I. till I see him across a jam up Mad River in th' spring, a drivin' his spikes into a log."

"'Hey,' I shouts, 'you P.I., what did you do with thet money y' stole off that monk?'

"With an angry expression, he shouts back, a-wavin' his pike pole at me, and down the river he goes a ridin' them logs like he was at one o' them ro-day-o's out west a ridin' broncs.

"But I never forgave P.I. for them two pair o' shoes I wore out that time a walkin' them blasted ties all the way to Lincoln."

**LUMBER AND LOG BOOK.**

Loading Logs on a Wagon—The Cut Explains Itself.

## Two for McMaster and One for Green

"Now mind you Alfred," continued Angus, "I don't hold P.I. Kelly had any right to a took that money off a poor man if he was only a ---- and a monk at that. Course it's alright if you put a bit over a rich feller now and again.

"And that reminds me of the time me and McKenzie was a cuttin' pulp for Green up on the upper Pemigiwasset.

"Mister Green, he sees me in Plymouth, and says he, 'Angus, I got a nice job for you until time to open camp, a cuttin' a bunch a pulp we didn't have time to get last winter, $2.00 a cord and a good scaler.'

"'All right,' says I, 'if I can find a man to work with, but don't you send up that pimply faced little son of a gun you had down to #2 last winter.'

"'No,' says Mister Green, 'I won't.'

"Now I never had no kick on Mister Green, he was as square a man as ever lived and took charge for the comp'ny, old Green, I mean. Well, I happens to meet Dan McKenzie that afternoon, and I asks him what he was doin'.

"'Not a thing', says he, 'but tryin' to figger out some scheme how t' fool my landlady.'

"'That's easy', says I. 'Marry her and you won't never have to do another lick o' work.'

"'Well,' says he, 'there's somethin' t' that, but she's death on good likker.'

"'Well', says I. 'I'll tell you how to fool her then — skip out this afternoon and take the train with me for the upper Pemigiwasset. Green's got the best little job he says a man ever had. Cut four cord a day, two dollars a cord, and throw into the river.'

"Well that struck McKenzie about right, so up he and I start for the upper Pemi with a few dollars advance from Mister Green in our pocket for grub.

"Well when we got there and see the bunch o' scrub spruce Green was seein' us two old-timers pullin' down at four cord a day — I coulda swore till the sky was grey. Well, McKenzie just give me one wicked look, but he knew he wuz stuck so he didn't say nothin'. Well, I had to laugh when I come to think of it.

'I guess Mister Green's a payin' us back, McKenzie,' says I, 'fer that big payroll back last summer when we plugged the holes in them old growth pine. By !!! he's got us this time, but he better watch out.'

"Well we sets to work like two sons a' guns for one week and the best we could do working half the night too was three cord a day. That want so bad but we wuz losin' sleep and not takin' time to cook decent grub.

"So I calls up the central office.

"'Mister Green', says I, where's that scaler? We ain't got no room to pile.'

'Why didn't you start pilin' further back?' says he. 'Well', says I, 'we got twenty cord and we ain't got warmed up yet. Been takin' it easy to start, jest a-workin' a few hours a day.'

'Well', says he, 'I'll be right up.'

'Well up he comes with that young kid and they looks at the wood kind of surprised like and they goes onto the lot. Then Green he comes back and the scaler he starts a scalin' th' wood we had piled on the bank.

'Didn't I tell you, Mister Green', I says, 'I won't take no figgers from that kid?'

'Now, Angus, mebbe you did, but the other scaler was sick and I thought —

'Now Mister Green I mean it', says I.' 'Me and McKenzie ain't agoin' t' cut no more wood if HE scales this wood.'

"Well, Green knows he's stuck. He couldn't get that wood cut for $2.00 by nobody else, and he says, 'All right, I'll send up Marsh.'

"Well, next day up comes a little feller 'bout as big as my thumb with big goggles on.

'Well, sonny,' says I, 'did your Ma let you come out today?' Says he, kind o' scared like, 'Mister Green says as how I'm to scale your wood.'

"Well we took him down to the river and when we sees that he was only marking the top sticks, I gives McKenzie th' wink. We thanks him very much and tells him that we'll be ready by Thursday next again.

"Well when two o'clock comes next day we quit and went into camp. By nine o'clock the next mornin' after th' best sleep we'd had for a week, we was out a-pitchin' that wood into th' river to make room for another pile. Well I tells McKenzie just to pitch the top sticks that was marked and we done it and then we goes to chopping again and rollin' down and sawin' up, and we piles right on top of the old sticks that wasn't marked. Well we does that fer a week just workin' enough to make an average of four cord a day what Green had told us we could do.

"Thursday next we calls up Mister Green and says I, 'Where's that scaler Mister Green? McKenzie and me's been workin' like two sons a' guns all this week but we got them four cord you told me was an easy job for two men the likes of us, and say Mister Green we ain't et or slept this week,' and I nudges McKenzie in the stummick so he'd stop a laughin' and Mister Green'd hear him thru the telephone. 'All right Angus,' answers Mister Green, 'I'll send Mr. Marsh right up."

"Well up comes Mr. Marsh the next mornin', and he scales that wood and he never smell no rat, no sir, and he just marks th' top sticks same as before. Then back he trots to Mr. Green.

"Into the river we throws the wood again leavin' the bottom half, and next day after a good sleep and a good breakfast on tenderloin and flapjacks, we goes to choppin' again and calls up the scaler the next week. Up he comes, back he trots, and into the river goes the sticks that is marked — McKenzie fairly a-busting himself for laffin' and a hollerin', 'There she goes, two for McMaster and one for Green,' a tossin' them sticks in like they was chicken bones.

"And there she went, 'two for McMaster and one for Green', till I says to McKenzie,

"How're y' feelin', Dan? We've got quite a boodie here, and Green, he ain't no fool. We've cut all the count called fer anyways. I guess we better hit the offices 'fore Mister Green shows up here and begins countin' stumps.

"So down we goes that afternoon and asks for our checks. 'All done, boys?' says Mister Green. 'That little piece figgered out to a T just what I figgered.'

'But you thought how it'd take quite a bit longer'n you figgered, now didn't you Mister Green?' says I nudgin' McKenzie to keep the fool from laffin'.

'You never thought as how we two old timers'd average that four cord a day you was telling me we'd get that easy.

'Well Angus, now to tell the truth I did put it a bit strong, maybe, but you don't look as if it'd hurt you a bit. I never see you looking so fat.'

'That's nudging that fool of a McKenzie t' keep him from laffin'.

'Now you go up there t'morrer, Mister Green, and see if you ever see a neater job done,' says I.

And I takes the checks, in three figgers too and hands his to Dan, then we shakes hands all around and I turns to go out, 'And,' says I, 'you just remember Mister Green it's two fer McMaster and one fer Green.' Then we both of us bust out laffin' and Mister Green he laughs too just to be sociable, but when we turns round to look again as we go down the road, there's Mister Green a lookin' after us and a scratchin' his head a much as to say,

'Now what did the old cuss mean, two fer McMaster and one fer Green. I'll be goin' right up there tomorrer morning.'

"Well I see Mister Green in Plymouth a while after and he's makin' a bee line fer me.

"Well, you old son of a gun,' says he. 'Two for McMaster and one for Green!'

"I'd say quicker, 'Four for McMaster and one for Green,' and he fit to bust laff in.

'And now Angus,' says he, 'was that scaler in it?'

'Honest to God' Mister Green,' I says, 'that scaler's as innocent as a newborn babe and as fit to be in the woods. But we just thought you needed a lesson crackin' up that job to an old timer like me and getting' us up there when you knowed we was busted, and I tell you I never laffed so much in my life, heavin' that pulp into the river with McKenzie a shoutin', **There she goes, two for McMaster and one for Green.'**

"'Well, serves me right,' says Mister Green. 'A-sending that Sunday School kid into the woods in the company of the likes of you and Dan McKenzie.'"


### Mr. Dale's Ideas on Religion

"That must have been the winter I was in Groton," commented Mr. Dale. "I was up there workin' for Tobey and boardin' at the minister's. That is, he called himself a minister tho I never noticed he ever did any preachin' save at home. He had a farm up there and two pigs and two old horses. Said he was retired but I guessed more likely they'd a fired him. Well I felt sorry for them two old horses and after I put my team in I used to feed and water 'em up when they hadn't had no water all day, and clean 'em up a bit while the 'minister' he was in the settin' room holdin' family prayers. I'd come in later, and one night he called me down for not takin' more int'rest in religion and tendin' prayers.

"I was mad clear thru that night for one o' them horses so thirsty he'd a drunk the tank dry if I'd let him and I knew that old hypocrite hadn't been near him all day.

"'Religion,' says I, 'you call that religion?' And I pointed at the hymn books and toggers he had in there. 'I'll tell you what religion is,' says I.

'Well perhaps you are better able than I, Mr. Dale,' says he.

## A Convenient Wood Holder.

It consists simply of a portion of a hollow log sawed off squarely, about one foot long and placed on one end for holding the wood while it is being split into small sticks. Such a contrivance saves labor, as it keeps the sticks erect, so that a workman may swing his axe freely; also saves time in picking up and adjusting the billets to be split. To prevent the numerous blows in one place from splitting such a holder, pin a half-round stick on the upper end, against which the axe may strike.

"'Yes sir, I'll tell you what religion is,' I repeats, 'and I don't pretend to be no minister of the gospel,' says I. 'Religion is goin' out and feeding them old half-starved horses when they need it and seein' they have a drink at night when they're thirsty when they ain't had a drop all day. That's religion Mr. Minister,' says I, 'and I'm going to find me another boardin' place with a little more of that kind of religion in it!'

"Yes sirree, it certainly makes me mad to see how some folks treat dumb animals, think they ain't got no feelins,' continued Mr. Dale.

"There was Death's Head Kelly — no more feelins' for a horse than it was a stone. You remember Death's Head Kelly, Angus?"

"The feller that drove the bays over at Thornton and sluiced 'em down the slide?"

"That's the same, only I knew him when he had a head of yaller curls. He was that vain of them and soft on the gals. They fell for him, too, like a lot of pullets for the only rooster in the barnyard. They called him Dude Kelly then.

"Dude Kelly," repeated Angus, "why I knowed him but I don't recollect ever seeing him for a long time. A cruel bugger then, struck the best horse in the camp one winter he worked for me. Claimed he was balky. Now you can hit a balky horse just right so to fell him and when he comes to, he'll forget he ever was balky. But Dude Kelly hit him once too often and they drew the horse to the boneyard. Well the old man was mad — just overloaded the brute! He was a beauty too. I kicked Kelly clean down the road and the old man never said a word. Never heard of him after. Never thought them two was the same. Death's head Kelly! Well, I'll be! You don't say. Couldn't of been very old when he was in Thornton either.

"Didn't you ever hear how he lost them curls, Angus? Well I'll be licked. But I guess I did hear as how he steered clear of your diggin' but I never thought much about it." Mr. Dale hitched his trousers up to loosen the tension.

He had had an extra piece of apple dumpling that evening, and settled himself still more comfortably in the only available armchair.

### Death's Head Kelly

"It happened over in Camp #3 in Woodstock, you know, Angus, just before they sold out to Beebe River? Death's Head Kelly, only he was Dude Kelly then, and the other teamsters was a-standin' around the stable door. Been a nice warm spell, but that night come off bitter cold. You could see the breath risin' like a frost off th' horses' backs. They'd just been put in from their last trips. Nice teams in those days Angus. Remember that big pair o' bays Luke Reagan drove at Thornton? Well they was there in camp that winter too. But as I was sayin' we was standing around waiting to water the horses and we got to talkin' about miracles. Dude Kelly he was scoffin' like, said there was no such thing as a miracle. Now, there was a stable boy there, used to build the fires and clean the shack and make beds, red headed feller, kinda simple at that. Mike was crazy 'bout animals though and he'd caught a robin that the cat had taken its leg and Mike had built a cage out of hay wire and hung it behind the horses and daytimes it was warm he'd take it out in the sun.

"Well 'fore we knew it Dude Kelly had took that robin out o' the cage.

'Huh' says he, 'I'll show you there ain't no such thing as miracles', and there he was 'fore we had a chance to speak, we was that dumbfounded, astrippin' the feathers off that little bird's back, all but his head and wings.

"'There', says he, holding the poor little bird by its wings all naked, 'fly to Jesus', says he, 'and show 'em there's a miracle'. And into that stingin' air that little bird flew. We just stood there rooted as if we could watch it a-freezin' to death up in that stingin' air. Nobody doin' a thing to that beast a-leerin' up into the sky.

"But just that minute Mike the stable boy had come around th' corner in time t' see what that brute had done. He made a roar like a stuck pig and 'fore we had time to say Jack Robinson he'd drove the dung fork into Dude Kelly's shin and he'd a killed him then and there if we hadn't come to our senses and druv him screechin' and blubberin' into the stable. And there he stood and cursed that dirty brute a-cringin' there, cursed him till a man's blood ran cold. And he screeched,

'May ye be bare as a grinnin' death's head 'fore morning.'

"Well, by gum, we was a eating at the table next morning. The Dude comes in, goes over to th' glass to admire hisself, ain't none of us looking up from the grub. We knew the old man had wind o' somethin' by the way he was lookin' and we all knew the old man was a tiger when he was mad. Then all a sudden,

'My God,' the Dude cries.

We all looks up just as we see him jump, makin' a grab for Mike as he was passin' the grub.

'G-- D---- you and your miracles', he cries, frothing at the mouth. The old man jumps up like a tiger.

'G--D---- you dirty brute', he says, and grabs him by the shirt and throws him out thru the door into that black cold.

"'There', says he, 'you miserable cur. Stay out there and freeze like the little bird ye murdered.'

"For it want no head of thick yaller curls Dude Kelly had saw when he went to that mirror to admire hisself, but a dome as bare as a grinnin' death's head.

"Well, I'll be darned!' ejaculated Angus, "I'll bet the dirty sinner didn't lay starboard of no more pretty gals for a spell after that."

## Black Martin

"And if that don't make you believe in miracles, Angus, listen to this I'll be telling you."

To be sure Angus was as ardent a believer in miracles as Mr. Dale, so I was convinced that Alfred knew I was listening in and was saying this wholly for my benefit.

"You remember Black Martin, that river boss up at Errol, don't you Angus?"

"Sure, the boss that was drowned in his cellar?" replied Angus.

"The very same, Angus, but did ye ever hear tell how he was drowned?"

"Like any damn fool, I s'pose," replied Angus. "I got no use for a man who'd kill himself for a woman. Just clear out I say, and Martin one of the best river bosses I ever see, too."

"No sir, Angus, you listen and I'll tell you," and Mr. Dale moved his chair just a bit so he could see thru the crack in the door if his unseen audience was still there.

"He was a good feller, was Black Martin. They called him Black Martin because he was part Indian and most as dark as a mulatto, straight black hair like iron wire — my God, he was strong, like a rock maple stump. Well, Black Martin had a wife, you remember her Angus, a little French woman, no bigger'n a peanut and white — you'd a thot Black Martin'd a took to a huskier dame than that. But t'was sayin', and you know, Angus, there ain't any man fool enough to let a woman think he's got a soul wuth savin' even if he ain't. Well as I was sayin' she got him anyway and then she started the savin' business. Well did ye ever see a little white thing begin workin' on a man? She just fastened that "want to save someone" feelin' on Black Martin — you know how they do it, make a man feel he's a sinner, a lost soul they call it, then they begin to work on him and when he comes home after a bit of celebration, just look white and pained and accusin' like — oh, I know they don't have to say a thing," and Mr. Dale sighed reminiscently, "till a man wants to take that scrawny neck in his hands and choke and choke. I know. Angus, and there ain't no milder man than meself. Well that's the way Black Martin felt.

'God, Alfred,' he says to me, 'did you ever see water drip on a piece of stone where it had been drippin' for a thousand years, like them holes up the river where the perfessor showed me till the stones ground out as if a grindstone had worked it? Well that's what my woman's doin' t'me. I ain't a bad feller when I'm sober but by God, Alfred', he says, 'you'd think I was the worst brute in the country. God, Alfred, a good woman's the worst devil the Lord ever invented sometime. Take 'em without no religion, if I could do it over again anyway.'

"Well, Black Martin got so he was scared stiff o' that accusin' spirit he had for a woman stid o' flesh and blood wife. And he'd take a swig every time afore he went home or he'd get roaring drunk over at Cotey's.

"Well one night he got drunker'n ever till he was mad crazy and into the house he lumbered. His wife had a crucifix over by the doorway, the Lord a-hangin' nailed to the cross. When she sees Black Martin in a-kickin' the cat halfway across th' floor, she just give him one long sorrowful look and down she plumps a-opposite that crucifix and starts prayin' fer his lost soul.

"Well, by the Lord, it just must have move over Martin till it drove him blind with fury, and up he grabs an axe and stood there a-glaring at that crucifix. I guess she thought he was going to hit that crucifix and she just knelt there and screamed,

'No Martin, not that, not that sacrilege on your soul, hit me instead!'

'Curse you and your religion,' yells he, and crashed the axe straight thru the Lord Jesus there on the crucifix.

'Look, look!' cries she, 'the dear Lord bleeds. Oh Martin, ye have murdered the Lord Jesus.' And by the Virgin Mary, Angus, down from the Lord where the axe had cut, the blood was drippin', drop by drop."

Alfred looked at Angus as if unsure of his ability to believe in miracles.

Angus said, "It ain't for me *not* to believe in miracles. Wasn't it a robin as brought me the message of me old mother's death last spring? One hundred years old she was, just think o' that. They don't breed such stock these days. And t'was this pair of socks she knit with her own hands, I'm telling you, and no glasses neither. A pink face she had, with flesh as soft as a baby's.

"Yes sir, t'was a robin brought me a word. The spring days had brought 'em out and then the mornings got frosty, and one morning I heard a pecking at me winder and I looks up and there's a robin a-trying to come in. I don't think much of it, but next morning the same little bird tries to wake me up again. T'is trying to say something, I said to myself and it flies away. Next morning it wakes me again a-peckin' at me winderpane.

"I tell Miss Park about it when she came over. T'is sure I tell her, the robin was trying to get me word from someone that died. She just laughs and says you must be superstitious she calls it. Well, the next week she comes bringing me a letter and sure enough, me dear old mother had took with a shock the very mornin' the robin first come a-rappin' at me winder and she died the same time the little robin come to get me her message that third mornin'.

"I was the youngest of the family of nine boys and my sister Hetty is now nearing eighty. And Hetty she writes that mother kept talking as she come to about her baby 'Angus', that's me you know. Though there was another Angus too, me oldest brother. We sort of run out of names when I came so me dear old mother just start in again. She called me young Angus y'know so we wouldn't get mixed up, but I was her favorite always. Many's the time I've held her yarn while she knitted the socks for the whole family. And that robin was tryin' to tell me mother's message them three mornin's he come a-rappin' at me winder."

# PART IV

Ruth A. Park:

Letters and Other Communications

Vassar College Library
Poughkeepsie, N.Y.

Photo from the 1906 Vassar College yearbook.
The caption next to Ruth's picture reads:
"There ain't a face but what she's shook her fist in."
*Vassar College Special Collections*

# LETTERS & OTHER COMMUNICATIONS

Ruth graduated from Vassar College in 1906, and the following are letters she wrote to the Vassar Alumnae Bulletin for the reunion year write-ups.

Vassar Alumnae Bulletin:
1916 Tenth Reunion
Park, Ruth A.
Beaver Meadows Farm, Warren, New Hampshire.

Why is it that college, while enriching the background of my aesthetic life, helping me in the power of general judgments and sense of values, should have been for me a dead loss and actual hindrance in fitting me for the struggle for existence? For I gained in college none of the power for continued application and attention to detail that is so necessary to the business of life. Intellectually it helped me, but not in the power of execution.

The ten years since college I have spent in gaining what college did not give me even in part.

Three years after college, finding that I would have to be in this part of the world for a few years anyway, I started farming, hating to see this farm lying waste.

One morning three years after that, when I had really a good start, a barn full of stock and had started in logging the woods on the farm, I saw in two hours all the work of those three years just a charred mass; yet not all the work, for the stock was saved, and I had the dreams and knowledge of the problems of farming, new ideas as to marketing especially, which we had gone out of the general course to put into effect. It is as vivid as yesterday: the great mass of flame eating through the long set of buildings, the furniture all about the place, bureau drawers sprawling open, the farmers' wives poking into everything to see what was what, the little children playing house as if it were a vacation day, the farmers standing about discussing the insurance and spitting tobacco juice, the lower end of the beautiful meadow covered with cattle, pigs, calves and horses grazing contentedly, and in the midst of it all my father sitting in a big armchair, crippled with rheumatism at the time, watching it all, while the tongues of flame, crackling and roaring, ate their way through the woods. Then came a new experience; we moved down into an old house where the help had been kept, paper half off, plaster patched on here and there, floors that see-sawed.

We lived in two rooms by ourselves; on the other side of the partitions were the lumberjacks' quarters. Outside were the old hay barns, the cattle tied in makeshift stanchions, open drains and the pigs reigning supreme in a hurriedly made pen in the front yard! Then came pandemonium, the logs to be got in, the cows to be milked, the barn to house hay and stock to be built and the old ones to be torn down as fast as the new one went up. For a year and a half I pioneered it until the woods were down. Those lumberjacks I bossed and scolded, kept sober nearly all the time, and then paid off. But it was worth it all, opening up a new side of things and a pleasure in the woods work that the human element could not dim. I have always loved it; the good smell of the pines, the sound of the axes and the crash of falling trees, the creak of the big teams swinging down the mountain side, and through it all an air like wine and a sparkling sunlight. It is intoxicating. One feels the perfect rhythm in one's body.

Then again I found myself ahead of the game, barns full of stock and fodder, a head beginning to get full of knowledge and ready for more work in the different problems of the farm; another winter of logging after a year free from it, and above all ten years back of me of training that had begun to show results.

Then I patented a new method of exit from the barn, chuting the chutes, but this time a hay chute, necessitating a fall of thirty feet, cast after a novel operation; and finally the knowledge of what friendship really means when Lois Warren Shaw rescued me from the hospital and took me to her home for the last six weeks in my cast and the final recovery.

Now here I am again on the farm, in two months to be as good as new. I hope the next ten years will see me able to succeed in the work itself, and besides connect it with the world outside the farm, helping solve some of the problems that need solution. As far as offspring is concerned, I can offer only bossy calves, unless as nieces you allow Lois Shaw's four children.

Vassar Alumnae Bulletin:
1926 Twentieth Reunion
Park, Ruth A.
Beaver Meadows Farm, Warren, N.H.

Since our first reunion I have spent my time turning many a disastrous somersault, the while madly clutching at the stars — at the same time pursued by furies that seemed to have been sent by fate in the shape of fire and death, that alternately stalk me. Seven years after the first fire, and one month after my father's death, came another fire — wiping out every vestige of our rebuilt home, except the barn and stock. I was away at the time at the lumber camp below when I received word that the farm was burning down. As I reached the ridge at the head of my lovely valley I could see nothing but flame. It was one of those steel-blue January nights of a stinging cold far below zero, but as I reached that ridge a mile from the farm, the air was hot against my cheek. The flames were devouring the whole mass of buildings in a driving wind. It was raging fury the fates had chosen this time. Our efforts could amount to nothing. After this, five years ago, I deliberately cut out the farm for two years. I felt almost as if the fates were bound against me and as if my old background of family and things, with whatever of beauty and leisure they had meant, were gone; so for two years I put in my time at the lumber camp, aiming only to retrieve a part of that disastrous war experience, living with the utmost bareness and frugality, but not devoid of the stimulus of natural beauty, which reacted only on the senses, however.

I moved into a little shack up against the edge of the woods. Never will I forget the gleaming sunsets, the incessant chirping of the first crickets on the still evenings, and through it all the wood thrushes and veerys, bringing me in touch with all the loveliness of the ages. With this song still in my ears I went back to the farm, resolved to find a way to farm and still keep well rounded my love for beauty, whatever form it might take in a farmer's life.

And here I have been for three years, rebuilding while I farmed as I could. The old farmers regard me as a riddle, but the children accept me on my own ground. I'm still tackling the problem, but seem to see a glimmer ahead.

Tomorrow, I shall see the first delicate haze of the green and red of the maple and poplar buds beginning to spread like a veil over the pasture side.

Yes, I have something worth it all.

Vassar Alumnae Bulletin:
1931 Twenty Fifth Reunion
Park, Ruth A.
Warren, New Hampshire.
Maskoma Lodge, Fairyland,
Dorchester, New Hampshire. (Winter address.)

Above address is mine for the winter, for here I am twelve miles back in the woods in a gleaming icicle-hung country, a veritable land of magic.

My life is still a part of the land, picking out ash or veneer logs in winter so that the rest of the land may not be turned over to the tax collector, and in summer alternately harvesting crops from the farm, and then crossing the divide into this country and wresting secrets of the thirties (1830) from the hidden roads through these woods to the entertainment of twelve year olds and some sixty year olds who still retain their youth.

For "Auntie" Ruth surely does enjoy spinning tales of bygone days, and is too often amazed to find that the twelve year olds take them with utmost seriousness; in fact she surprised one lot of young things laden with pickaxes and spades en route for Maskoma's cave one afternoon, and another night around midnight another group hidden on the Phantom Buggy trail awaiting Silas Burnham and Willy Davenport with the old white horse and Phantom Buggy! For in summer Maskoma Lodge is a summer camp filled with grown-ups and children, run by my sister with my very irregular help.

Farming still pulls along apace with the times, much pleasure, but not a great income. Work with the village boys on the farm in summer constitutes another real pleasure besides companionship with the camp children. I find I like children much better than grown-ups!

Winter is not so starving mentally from lack of association after all, for there are four radii with Maskoma Lodge as a center. On the end of one is a farm family, five miles from nowhere, that is comprised of two of the best informed minds on the municipal and state government (coupled with a dry humor) with which I ever came in contact. At the end of another radius of some six miles is another family, the father of the family, although a stone mason and small poultryman, possessing a semi-radical point of view, a single taxer of fundamental information, and the wife one of the most mellow personalities I ever had the pleasure of enjoying. And at the other end of the opposite radius is another family, now a successful poultryman of advanced years who hailed originally from what is now an abandoned farm of this back land, spent his middle years all over the states, then returned, broken in health and finances, who too has this capacity for original thinking with a wife thoroughly his equal. And at the end of the other radius is a group of people with the same capacity, hailing from the city, fighting disease and winning out, now small farmers, but with minds as active as bees all along the line of conduct and government of living.

Curious, all these four families, about the same distance apart, more or less isolated, seem to possess far more active mentalities, richer minds than equivalent groups that one meets in more thickly settled communities. Although in no one of these cases is there any particular love for beauty except as expressed in order and love of nature, probably this is from lack of leisure and means, such hard physical labour being necessary to wrest the necessary income. But the city man is mistaken if he thinks their minds are rendered less acute or original thereby, or their zest for living lessened.

An interesting life, and I am finding a certain contentment. I find I miss these points of contact and natural beauty when away from them long. Here's to seeing what the years will bring after I succeed in allotting a certain amount of leisure sure to come when I can sell off a part of our land, almost arranged for when bad times came around the corner last October — the chance sure to come again soon I hope.


Vassar Alumnae Bulletin:
1946 Fortieth Reunion
PARK, RUTH A.
Maskoma Lodge, Lyme Center, New Hampshire

Ever since the hurricane September, 1938, I have been struggling to retrieve what I lost then. I sat in the window on top of this divide and watched the wind and rain for one hour destroy what I had finally brought to a paying condition with a rosy future ahead. I struggled with jobbers for four years to sell what the stumpage buyers had given up. My feelings were mingled, relief mingled with disappointment, struggling with elation. For I was going to do something I had always wanted to do, get a close-up of industrial conditions in the world outside.

I went into a huge cargo-plane factory outside Philadelphia in the country training for a welding supervisor. But I made up my mind I would be a regular welder, in this way getting closer to the workers.

It was some exchange, my old home up here beside the lake and the mountain, awakening to the creaking of sleds on snow, the rattling chains, the shouts of teamsters to their horses, the booming of the lake as the ice cracked, like a huge giant belching and groaning beneath it as he turned in his sleep. All these familiar sounds were changed to the roar of armored trucks racing along the main boulevard, the reverberations of the main-line freights. My room was like a cell in an old stone schoolhouse. My only neighbors were the Polish couple and Mr. Smilski the owner. But Mr. Smilski was only a voice as he boomed good-morning to the Polish couple across the way, then out with a bang of doors. I believe he kept a taproom in the city.

I was not looking for a laboring man's Utopia, but from all I had read in the magazines and heard from Roosevelt over the radio about the wonderful effort of industry and labor in the war effort, I was ready to make allowances for war conditions, shortage of labor, the experimental nature of this particular project. I did not expect any wonderful physical conditions. Factories I had thought were dingy places at best. And I was looking for hard work and lots of it.

I certainly changed my mind that first morning about factories being dingy places. I never dreamt a factory could be beautiful but this huge place was, one third of a mile long, airy, artificially lighted with glass roof and sides, spacious restrooms, canteens, everything one could imagine for the physical comfort of employees, except chairs. But did I find patriotism? If it was there it kept well hidden.

The assumption that management's job is done when they furnish good pay, good surroundings and work is the cause for such an attitude of mind as I came across in my factory experience. The reason for this mental attitude must be found in our education. I felt more and more what a seedbed was in the preparation for the radical, the power-seeking and the demagogic elements. There was no attempt except in isolated cases to get at the core of the matter in our religious and general educational methods.

And I wondered if the intense capacity on the part of the

individual worker for personal kindness would be strong enough to neutralize their mass hatred and antagonisms when whipped into action by the unscrupulous leader. For I met that kindness everywhere.

Now I am back again with my mountain and my lake. Hunters are all around me. I feast on an unearthly beauty of sunset coloring every evening. I am as warm or as cool as I please. My one companion is a big black Newfoundland, except for the chance meetings with hunters. The first snow has come. I am trying out again on high piece wages a little logging. I may go back to Philadelphia, New York, or Boston when the depth of snow stops trucking.

Best wishes to all. I feel intensely active and hope this is true of all of you.

Vassar Alumnae Bulletin :
1956 Fiftieth Reunion
PARK, RUTH, A.
Dorchester Woods, Lyme Center, New Hampshire

I'm still logging this 5,000 acre tract lying between three skylines and bordered by the sunsets and the Dismal Swamp country on the west. It's a country of hills and woods and swamps and ponds, with the lake and mountain on our immediate borders; with deer and mink and the inevitable rabbit and bobcat and bear; and now overpopulated with the over-industrious beaver — whom the game warden lays at my door, as I got him to bring up the first pair and let them out on our lake edge.

I love the life here, even in the summers when I am surrounded by the grown ups from my sister's summer camp, and their children — particularly their children. To be sure, after the camp closes, it is a life of solitude; but I am inclined to believe with Thoreau that there is nothing so companionable as solitude; unless it is the sociability of those summer children, invariably responsive to the beauty of animal and bird life that we encounter on our rambles.

I'll admit that, once in a while, I have a guilty feeling when I think of the activity of such 1906ers as Cary Baynes in her intellectual accomplishment. Then again I sometimes have a very pleasing sense of self-justification when I think of those summer children or when, perhaps in a railroad station, one of the many young men who have worked for me comes up and says "Don't you remember me? I'll never forget how kind you were years ago when you helped me to the job that was the making of me." That sort of thing sets my feet on a path of air!

My financial difficulties are all behind me now and I can even see ahead of me a possible trip to the Gobi Desert — always one of my archaeological goals. But it has taken 50 years of solitude and simple living to come out on top. But it has been quick sailing since I broke with the Sherman Adamses of the business world — the veneer companies with whom I had been working until I pulled up stakes, as I wrote you in the last Bulletin, and went into the factories in wartime. When I returned, I started logging again in a very small way, with entirely new connections — connections whose faith in my ability to carry through my commitments were in great part responsible for my success.

I still follow the land agents running the lines along the property, and shall do so as long as my hips hold out, but I have given up trapping beaver ever since I dug post holes through 15 inches of ice and caught only three beaver! The otter will catch them, though they catch the trout as well, and otter are such graceful, intelligent creatures that I refuse to trap them.

I think that it is the cracker barrel in the village store and the summer children that I must depend upon for my future socializing contacts, but will even the cracker barrel remain for a clearing house of political ideas? You can feel, even now, the intolerance and suspicion, bred by the attitude of the administration toward the very thing they are trying to destroy. We need a man of high intellectual caliber, like Stevenson, to weigh each side and make sure that more freedoms are not lost than gained.

## And the "lumbering lady of Lyme" goes on!

### Letter to Richard, sometime in the 1950s

Dear Dick,

Armistice Day — . . . McCarthyism to me is as dangerous as Communism. Neither relies on truth to gain its ends. I am all done with Governor Sherman Adams for signing that bill outlawing the Communists and anyone joining a so-called subversive organization in N.H. I always knew he had no vision even though a good administrator and that is also why I fear Eisenhower.

I am having a delightful time all alone here . . . hunting in the woods for another bit of cutting to carry thru up to the time of frozen swamps in which localities I can still find spots of two thousand each that one can afford to get on these high prices. I found one up beyond Mudgetts by slabbing the hill on P.Y. Co [Parker Young Company] land [—] a hard pull for a scoot however. Later I went up Black Brook by slipping down the trail at Bailey Cabin to where I knew there was a ravine of venner [veneer] then tried to see if I could pull it out above the old road along Black Brook — a long way and I saw its impossibility after going 100 rods, too much cutting and harder than the P.Y. road at Mudgetts — lots of blow-down also. So I cut down to the old road and went along for a bit then hit water up to my knees with a frozen crust of ice that would not bare [bear] my weight. And just like this all along except for occasional high ground, and getting worse the nearer I got to the main road. Water flowed around the tops of my rubber boots and for 100 rods I walked right thru it smashing the ice as I made my path. I could not go on high ground or at least I did not fancy forcing myself thru the blowdown, the blowdown this time of last November, the one that took the telephone line down and the top off the Shore Cabin. I kept crossing paths the beavers had made to carry their timber thru the water to the series of dams they have been making, until finally I got out of the water and ice and muddy bottom onto the little east branch road leading to our road skidways — only to find that full of blowdown, too.

Well as I got on higher ground and looked over the Dismal Swamp way over to the line of thick growth on the opposite bank of the brook, it seemed like a good twenty rods wide all the way towards the mountain — a virtual country of desolation.

Likewise to continue on the subject of beaver activities, I went

Ruth, her sister Katherine, and their sister-in-law, Winifred Higgins Park.
*Park family collection*

up at 1 A.M. towards Mudgett Brook (I thought I heard some fishermen just at the end of fishing season.) As I went thru the Mudgett Bridge I heard a queer grunting sound like a lot of little old men sawing wood, their breath forced with their too great exertions. I stood there but saw nothing. The next morning I went up and found beaver tracks all around and to the right close to the bridge a beaver house started. Already I had noticed cuttings before where they had been building up the embankment. Now the whole side on each side of the bridge is flooded — and all in two short weeks. . . .

Now it occurs to me since hearing a radio talk in Canada of the reaping of wild rice way up in the North if I could get ducks in here by growing wild rice in these two beaver made swamps. A hunter told me ducks like beaver ponds. . . . I am also going to connect with a trapper down at the village and see if he will go halves with me on some trapping if we watch some traps for him.

Also I am having a lot of friends here this fall, deer friends; and I got more delight out of them than I do out of the other kind of dear friends. Every late afternoon I shake down the old apple tree just below my cabin. Then soon, out come the deer, four of them, two grownups and two small ones. I can stand in my window and watch them eat the apples. Not a sound would I hear even when they were coming. It is just as if they materialized at the stroke of a hidden wand — the cute things. All four of them stood and watched me one night when my door was open by chance — while I talked in a lilting way to them — can you imagine me talking in that fashion? Then I made a movement and away they went, their white tails bobbing like great fluffy feathers so large I could not even see their bodies, only the light rise and fall of those feathery plumes of white. And not a sound to betray their quick movements as they disappeared thru the gap in the wall down into the next field.

Well — enough said. I am listening to the Montreal church service before composing myself for the symphony at 2 P.M. I listen every Saturday to Montreal opera from two to three P.M. Caruso and Nordica in Puccini's *La Boheme*. No it wasn't *La Boheme*, that shows what a good musical memory I have. It was Farrar and Scotti in *La Boheme* and Juss Bjorling in *Manon Lescot*. Boy! Either Caruso or Bjorling for me, much preferable to Mario Lanza. Maybe because he comes on at 10 P.M. and I go to sleep on him each time. Too late hours for me after a day in the open chasing up timber. . .

               With love,               R. A. P.

# PART V
## Postlude

Ruth A. Park: A Factual Account of Her Life in a Cabin in the Hills, and Commentary of Those Who Knew Her

Vassar College Field Hockey Team, 1906.
Ruth, in upper left corner, played left wing.
*Vassar College Special Collections*

## Ruth A. Park and Women's Rights

While Ruth was learning to organize her financial priorities and juggling the responsibilities of the Park enterprises, people across the world were learning to organize politically to secure the right to vote for women. In the United States, the National Women's Party was formed, soon to be led by a petite Quaker woman named Alice Paul. By 1920, twenty-six nations had secured this right for their female population. The Isle of Man had done so as early as 1881, and New Zealand, by 1893. However, the United States, the champion of democracy and freedom, lagged behind its neighbors. Why?

During the Civil War, the suffrage of the black man became the priority. Women's suffrage would come on its heels, was the assumption. But it was not the case. When the war concluded, little did women's suffrage workers know they would face decades of political stalling.

Flat refusal of President Wilson and his cabinet to bring the issue to the consideration of the Congress in 1913 left the suffragettes no choice but to construct a political machine to obtain their objective. The National Women's Party was formed, and Alice Paul became its leader. She focused all efforts on the passage of the following amendment to the U.S. Constitution: "The right of citizens of the United States to vote shall not be denied or abridged by the United States or any state on account of sex."

The NWP's civil activism began with picketing the White House. It was impossible for President Wilson to leave or return home without seeing the suffragettes, often dressed in purple, white and gold, holding banners stating their views. After months of this intense embarrassment, orders were given to arrest the women. They were charged with "obstructing traffic" and sentenced to the prison at Occoquon, Virginia. Yet, the activists proved to be more stalwart than expected. The picketing continued. According to *Jailed for Freedom*, by Doris Stevens, who served on the executive committee of the NWP from 1917–1920, the suffragette movement won over the hearts of hitherto uninformed citizens, one day at a time.

At an extremely critical point, in February of 1919, with the media of the world watching, 22 women volunteered to picket in Boston when President Wilson returned from Europe. Two of the demonstrators were Vassar classmates of Ruth's: Elsie Hill and Lois Warren. Elsie was the daughter of a congressman, and had become a dedicated worker in settlement houses. Lois was the

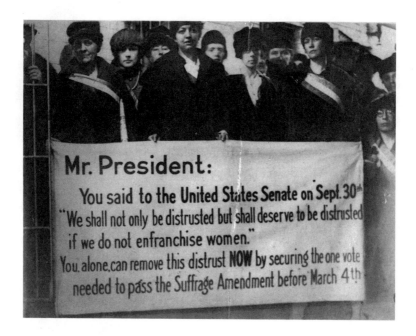

only daughter of Hattie and John Ebenezer Warren, of Westbrook, Maine. John was a popular mill manager of S.D. Warren paper company. (Lois is in the front row top right in the above photo.)

Ruth often visited Lois and her husband Winfield and their five daughters in the Manchester area. The two women joined the NWP, and remained lifelong friends, sharing ideas and feelings about the suffrage issue. In spirit, Ruth rallied with the Boston demonstrators, and as much as she wanted to be there in person, February is an active month for loggers, and Ruth's name does not appear on the list of those arrested. But Lois, her sister-in-law Pasha Warren, and Elsie were arrested and sentenced with the others to 8 days in the Charles Street jail. A media circus ensued. Embarrassed Boston officials heaved a sigh of relief when a mysterious E.H. Howe paid the fines, and the women were released. The event proved to be one of the turning points that secured the needed support from the legislature. On June 4th, 1919, the nineteenth amendment, submitted first in 1878 by Susan B. Anthony, was finally presented to the states for ratification, and by 1920 American women could legally vote in political elections.

The lumber queen's richest contribution to the movement manifested itself in the way she led her life; she proved by her own actions that women are equal to the challenges of running their own businesses, managing their finances, and surviving in a so-called man's world.

## CAMP MASKOMA

In efforts to reverse a decline in rural population, the Baker River communities began promoting summer tourism, as early as 1860, according to J.B. Hoyt in *The Baker River Towns*. The Victorian traveler, gladly escaping the humidity of eastern cities, sang the praises of New Hampshire mountains, lakes and streams.

Growing interest in physical fitness and women's rights resulted in summer camps for girls. By the 1920s, young ladies, once confined to nothing more strenuous than lawn tennis, were encouraged to develop recreational skills, like hiking, camping, tennis, boating, swimming, horseback riding. These new trends resulted in fashion changes — tight-fitting, elaborate gowns and beaded slippers gave way to the chemise and more practical footwear. The corset, restricting movement and breathing, reigned no longer!

Katherine forwarded the idea for the Parks to begin a girls' summer camp. According to her son Wistar Goodhue, Maskoma Lodge or Camp Maskoma, as it came to be called, operated from approximately 1917 to the 1950s, making it one of the longest lasting in the area. In later years, it evolved into a family camp. Maskoma Lodge was centered around the family home on the edge of Cummins Pond in Dorchester, the home where Ruth and her siblings had grown up in the 1890s.

Katherine and her husband, Fran Goodhue, lived in Germantown, outside Philadelphia, and many of the campers were from Pennsylvania Quaker families. Family friends also were invited, and Lois and Winfield's daughters, Mary and Rhoda, attended in the late 1920s.

In many ways, the camp was a family project. Katherine organized and administered; Ruth helped in promotion and planning, as well as preparing the site. Richard and his wife, Win, invested time and ideas as well. Financial investments mirrored Ruth's lumber business philosophy; remain lightly invested, and ready to change when circumstances arise; look within the house for what you need.

The old vine-covered farmhouse became the much-beloved "Lodge"; the barn transformed into riding stables. Ruth and her crews supplied the lumber and labor for screened log cabins which served as sleeping quarters. The Parks realized that the scenic beauty and leadership were the attraction, not expensive buildings and costly programs.

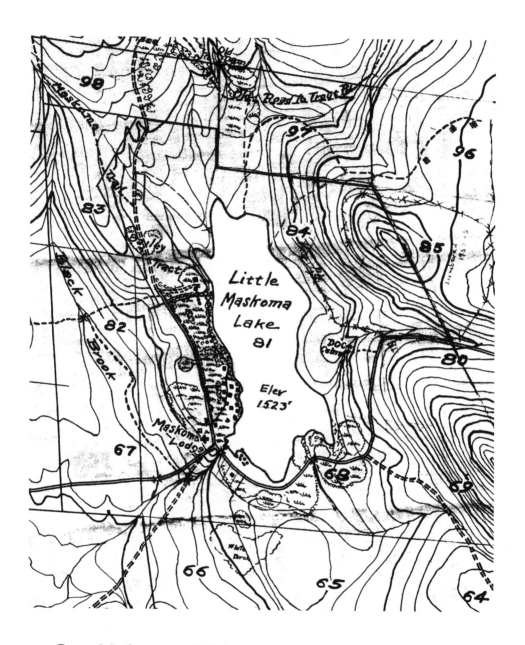

Camp Maskoma on *"Little Maskoma Lake,"* Dorchester, N.H.,
otherwise known as Cummins Pond.
Survey by Richard Park, 1929.
*Park family collection*

Maskoma Lodge, Dorchester, N.H., summer camp activities.
*Park family collection*

"The Old Dodge," Camp Maskoma.
*Park family collection*

A "Maskoma Girl" in camp uniform by the old well.
*Park family collection*

Camp Maskoma group on a camping trip.
Richard Park, back row center.
*Park family collection*

Mashoma Lodge's Moosilauke Song
    To tune of "What's the Reason"
Oh! we went to the farm
In the Dodge we bumped along
Then off for the top of Moosilauke we start
Along the ravine
And beside the mountain stream
Then from the ridge of Jobildunk depart
Up to the summit we puffed
And at Tip Top House we stuffed
On the pancakes of fame
And we'll <u>never</u> be the same!!!!
Oh we tried and we tried
To get more pancakes inside
'Till sliding down Hell's Highway nearly died!

So, they boldly promoted their highland paradise to the well-to-do and the educated, knowing that the simple, rustic life was a universal dream that stirred the soul of man. The promotion was successful, and created a rich tradition that brought families back, year after year.

Camp Maskoma's tantalizing brochure tells of a 160 acre pond called "Little Maskoma Lake" (Cummins Pond) under the shadow of Maskoma Mountain (Smarts Mountain). The 10,000 acre Park tract became "the enchanted playground which beckons all lovers of the mountains, woods and streams to come to the Greenwood and live a life as free as Robin Hood and his Merry Men."

The camp brochure was possibly written by Ruth. Descriptions of activities included

*Gypsy trips with pack and horse are great fun for the explorer who loves to roam over the countryside. A network of forgotten logging roads and old trails, following along deep-pooled streams, proves a never-ending source of joy to the horseback rider.*

Hiking was a strong feature of the camp, and Maskoma girls could be seen tramping the trails of central New Hampshire, including those in the White Mountain National Forest. Another popular activity was canoeing down the wide Connecticut River, making camp on the banks by night. For the less adventurous, or for rainy days, there was "craft work, homespun weaving, trailblazing and woodcraft."

The shining waters of the lake were enhanced with a diving platform, complete with spring boards, chutes, a high-dive, and a trapeze. Eventually, a sturdy, steep-pitched one-room log structure was built as an overseer's cabin. In later years, the fast flow of the brook leaving the lake at the spillway was harnessed to generate electricity for the Lodge. Hence, a pump house was built, as well as an ice-house and well-house. But the thrust of the experience was to lead an out door life, as much as possible. Under *Outfit*, the brochure reads:

*Old camping clothes are in vogue. Bathing suits, heavy sweaters, rubber blankets, raincoats, heavy shoes and tennis shoes are essential. Campers must send ahead parcel post, their own blankets, sheets, pillow cases, towel and a small pillow, if desired.*

Ruth helped open the camp buildings in the spring and secure them in the fall. When needed, she and the hired hands cut brush, did the haying, transported animals and supplies. Lyle Moody of Warren, who worked for Ruth in the 1930s, reports that several dairy cows from the Warren farm were trucked to the Dorchester Town House, unloaded, and herded up the Cummins Pond Road to the camp, where they would graze for the summer. This provided not only milk and butter for the campers, but cream for ice-cream making, a most-popular event. The Warren farm also supplied riding horses for the camp. Cabbages and other vegetables from the Warren farm supplemented what was grown in the camp garden.

Ruth volunteered to be a storyteller, and discovered she was good at it. According to Ruth's nephew, Richard's son, Joseph Dodge Park, who worked at the camp in 1939-40, "Auntie Ruth told wonderful ghost stories, around the fireplace at the Lodge, in her high-pitched, nasal, squeaky voice . . . tales like the 'Phantom-Buggy trail,' 'Porky-House with old Amos,' and 'the Lost Loggers.'"

Joe also reports "the theme of the camp when I was there seemed to be music. Katherine had a music counselor. The girls sang as they hiked, and did canoeing, or climbing Mount Moosilauke . . . songs like 'Maskoma Lodge,' 'Any Kind of Girl,' 'Moosilauke.' There was a camp song-book. At night, they would sing at the Lodge before bed, and then conclude with 'Now the Day is Over. . .' and a bugler on the hill would play taps. It was beautiful."

Libby Thayer, whose family lived in Dorchester, worked at the camp for three summers as a cook, riding instructor, and counselor for the children during Family Camp. She recalls Ruth would take the youngsters on walks through the woods to locate cellar holes of old farmhouses that were built during the pre-revolutionary period. She gave the campers information on the families that lived there so long ago . . . "Ruth loved kids, and she was a wonderful story-teller."

Katherine was responsible for maintaining the discipline at camp, and when romance blossomed between a handyman and a counselor, Katherine had to take a strong stand for the rules. The couple went out on the lake for an evening canoe ride, and didn't return until after taps. It was dark when they pulled up to the dock, where Katherine was waiting.

"Well," said the chagrined boy, "I guess I'd better be going."

"Yes," replied Katherine, "before dawn." So, the story goes, the young man left Maskoma with only 40 cents in his pocket. He hitchhiked to Bangor, where he found work on a lumber schooner.

Frugality was a prominent value of the Parks. Another young man, Alfred Balch, was hired by Katherine and Ruth to wash dishes at the camp for a week, until the cook arrived. A price of $25 plus room and board was agreed upon. However, when he received his check, it was only $12.50. They had deducted for "benefits, such as canoe rides, peaches, et cetera." Later on, Alfred and a friend of his were hired by Ruth to get the hay in before it rained.

"A dollar an hour all right?" she asked.

"Yes," they said. When the hay was in, she paid them a dollar an hour — 50 cents apiece!

Despite the strictness, many children found the Park sisters engaging and interesting, and learned something of the ideals on which Maskoma was founded:

*... love of silent places and the wild woods creatures, self-reliance on the open road, spontaneous and joyous initiative and cooperation in work and play.*

In the early 1920s, an electric company in Lebanon offered to buy the water rights to the pond. They needed backup supplies of water power during certain seasons of the year. This proposal drew into conflict two of Ruth's strongest values: the yearning to be debt-free and the love of the wilderness. True to style, she set out to discover all the facts she could before making a decision. Richard, a civil engineer with the US Army, produced a precise map of the lake, and calculated what the usage would do to the water level and shoreline. Ruth, over a period stretching into two or more years, acted as his research assistant, sending him pages and pages of measurements and data. In the end, the love of nature won out. The Parks did not sell the water rights.

Ruth wrote about Katherine's satisfaction in running the camp. "It is her real pleasure in life — a feeling of fulfillment. . . . The camp proved [to be the] size for a perfect operation for her abilities. . . . It was really a very well done piece of work. [She is] honest and simple in her motives . . . and the girls' side of camp furnishes her the energy and spiritual satisfaction she needs in any work."

Site of Camp Maskoma on Cummins Pond,
Dorchester, N.H.
Spring, 2001.

Ruth Park Cabin, Dorchester, N.H.
Cummins Pond and Smarts Mountain.
*Photos by Robert W. Averill*

**Ruth Park**

Ruth shunned being photographed. This is one of the rare pictures of her.
*Park family collection*

## PERSONAL PURSUITS

Richard and Katherine had growing families and active lives of their own, and Ruth, as a single person, struggled to find where she fit into the picture. It became apparent that relying on them for mental stimulus, and a feeling of belonging overtaxed the relationships, and she had to develop a life of her own. She learned to be more self-sufficient, discovering more within the reservoir of her own talents and interests. Reading occupied more and more of her free time, and also writing. Dartmouth students surprised her with the gift of a Delco radio. It was installed in the Lodge, the only building at Camp Maskoma with electric power. (Her own cabin had neither electricity nor plumbing.) The Delco was a source of great delight. One of the highlights of her week was listening to the radio broadcasts of opera on Saturday afternoons. And when time and money allowed, she traveled to Washington, D.C., the Northwest, Mexico, and beyond.

Yearning to free up additional time for her new interests, she began to look for ways to simplify the estate: In July of 1923, she wrote Richard: "I am letting Stebbins take back the West Rumney land . . . its value being about the same as the note. You will receive nothing."

In November, she wrote, "I am going to stick it out on the farm for two more years." But in fact she managed to run the dairy and poultry operation through the Great Depression. In 1936, Angus McMaster passed on, and Lyle Moody surmises that Ruth provided the headstone in the Warren cemetery, which reads:

**Angus McMaster 1870–1936**
**A loyal friend and faithful helper.**

On February 22, 1938, she wrote Richard and Katherine, "I've sold the farm for no money down. And moved into the Lodge [on Cummins Pond]."

For the next four decades, this mountain lake and its shores were Ruth's home. During much of that time, she lived essentially alone, miles away from other people. But in the summer, there were campers and family members, in the fall passing hunters, fishermen, and hikers to visit with, and winter brought the logging crews, and in later years, snowmobilers kept in touch with her. She kept a small crew of woodsmen active through many of those years, working the woods around the pond.

Henry Waldo, the woodlands buyer for Parker-Young Company from 1945-1970, purchased 4000 acres around Cummins Pond from the Park estate. He reports it was pretty well cut over by the time they bought it, but it rounded out the property they owned. He verifies Ruth's activities as a woodswoman.

"Yes, she could drive a team of eight horses, drive truck, and snowshoe well. She sold stumpage all the time I was acquainted with her. I recall worrying about her in later years, so I one day borrowed a snowmobile and took her a cache of groceries, but she had been plowed out and was gone. I never knew why that lady had such faith in me. She somehow felt I'd take better care of the land than others. We had a very cordial relationship."

Alfred Balch had permission from Ruth to fish in the pond, and one day with the temperature in the teens, he made his way up there to do some ice fishing. It was too blustery, so he packed up and found Ruth's tracks on the shore headed for Dorchester, some 4-5 miles away. He, too, was concerned for her, and followed the tracks. He met her coming back, having been down to board up a window on a cabin on Norris Brook. She told him to go back to her cabin, and start the fire, and they'd have a cup of tea and a chat.

"She really was the caretaker up there, keeping it wild and natural. It is a beautiful spot," says Alfred, who later, after Ruth passed on in 1980 in her ninety-fifth year, became the caretaker himself for the next owner. He could then fully appreciate what a responsibility it was.

"Rowdy, intoxicated fishermen were hard to handle, and often slobs when it came to trash. Was she happy? I always figured she was. She was doing what she wanted to do. She liked the independence."

In the winters, at times, Ruth would yield to the pleading of townspeople and officials who worried about her and find lodging for the winter in Lyme or Thetford. One family she stayed with was the Ulines in Lyme. The daughter, Isabel Pushee, recalls their unusual boarder.

"She'd bring old Sancho, the Newfoundland dog. He was a slobberer, but my mother liked him. Ruth would leave her good clothes with Mother, like a black persian lamb hat, and an expensive coat. She could dress up, but didn't very often. In fact, she'd wear high white socks and high top sneakers. And when it was real cold, a coon-skin coat. Whenever she would arrive, she'd bring a box of corn flakes.

"My husband, Roger, worked for Ruth in the early forties, before the War. He drove truck for her. When it was loaded, he'd take the logs to wherever she told him, Gale River, Conway. It was pulp. She ran a logging camp off to one side of the Cummins Pond Girls' Camp. All kinds of men there, some strangers. It took some woman to be the boss. Once my husband and Ruth had a run-in. He was driving pulp for her and she wanted him to go into her cabin and get her tweed coat and bring it to her in Lebanon. He brought her the coat the dogs slept on. She was . . . well, she had quite a disposition.

"She grew the most beautiful white phlox in her garden up there. Her secret, she told me, was horse manure. Those flowers grew so tall and had a nice scent. The hurricane of 38? Yes, she was involved in the cleanup." (The storm came right through the center of New Hampshire, leaving a path of destruction behind.) *The Canaan Gazette* reported that Ruth had a crew of 100 men, clearing the blown down trees, all tumbled together like thousands of match sticks. Isabel was skeptical.

"Not sure if that could be true. It would have cost her a fortune to feed that many."

During World War II, Ruth took a job for a time in the Philadelphia area, as a welder in an airplane cargo factory. She wanted to see if patriotism existed in the "real" world. What she did find were good, hard-working immigrants, men and women, holding up under the strains of war. Racism raised its ugly head in her jobsite. When layoffs began, Ruth learned of the management's design to fire the black workers before the white. Skin color, not skill, expertise, or seniority dominated in the decision-making. Any bubble of idealism she had safeguarded burst, and she spoke out about the injustice. Her actions amounted to little, and she left the factory before they could fire her.

The beauty of her upland kingdom comforted her on her return, more than ever before. She filled her free time writing her memoirs. Several people mention that Ruth was writing a book, and do not know if she ever completed it. But at this writing it has not appeared nor has her history of the early settlers in Dorchester. She explored the possibilities of having the *Lumberjack Tales* published, as well as her essay on the wartime factory experience. At some period, she traveled monthly to be a storyteller for crippled children. The works of Joseph Conrad, Antoine de St. Exupery, Hazel Hill, and Willa Cather, magazines like *The New Republic*, and opera and symphony occupied many hours of her solitude.

Her longevity, she would attribute, as many woodsmen did, to the long hours spent working out-of-doors. She enjoyed good health and approached her many interests with vigor and vitality.

When she passed on, in her ninety-fifth year, she was free of debt. She had transferred the remaining acres of the Park estate to Katherine's daughter and son, with special provisions for a piece to go to a grand-nephew and his friend. Liquid assets went to pay an account at Rich Brothers Grocery, and to her niece and nephew, with a few modest provisions. Her wish was for her body to be given for medical research.

But to many folks who grew to understand her and care for her, Auntie Ruth left a rich legacy in intangible blessings that have to do with beauty, family, and nature. Perhaps this heritage can be glimpsed in the poetry of a Unitarian hymn:

>*The blessings of the earth and sky*
>*Upon our friendly house do lie.*
>*The rightness of a master's art*
>*Has blessed with grace its every part.*
>*The warmth of many hands is strewn*
>*In human blessing on this stone.*
>
>*The wind upon the lakes and hills*
>*Performs its native rituals.*
>*The worship of our human toil*
>*Brings sacrament from sun and soil.*
>*With words and music, we, the earth,*
>*In nature's wonder seek our worth.*
>
>*Here we restore ancestral dreams*
>*Enshrined in floor and wall and beam,*
>*A monument wherein we build*
>*That their high purpose be fulfilled,*
>*A tool to help our children prove*
>*An earth of promise and of love.*

Ruth Ayer Park discovered a great deal of who she really was and what she was made of, in the lumber woods of New Hampshire. And wherever we work or live, we can draw courage and strength from her example by responding to a challenge: "I don't know that I can't do it. How will I know unless I try?"

To order books or prints from

# Moose Country Press

Call us
## TOLL FREE
## 1-800-34-MOOSE
(1-800-346-6673)

Or